Embracing Limits

Embracing Limits

A Radical and Necessary Approach to the Environmental Crisis

KEITH AKERS

ISBN 978-0945528-02-9
Library of Congress Control Number: 2022921365

Includes bibliographical information and index.

Global environmental change–Social aspects.
Sustainability.
Nature–Effect of human beings on.
Technology and civilization.

BISG categories:
SCI075000   SCIENCE / Philosophy & Social Aspects
SCI026000   SCIENCE / Environmental Science
BUS099000   BUSINESS & ECONOMICS / Environmental Economics

Earth Animal Trust
P. O. Box 11240
Englewood, Colorado 80151

Printing 1

For Kate

Dearest and sweetest, love of my life

In order to save our planet Earth,
we must have a collective awakening.
Individual awakening is not enough.

— Thich Nhat Hanh

CONTENTS

INTRODUCTION

Numerous environmental problems now threaten industrial civilization. Climate change is the best known, but more wait in the wings: peak oil, mass extinctions, water shortages, overgrazing by cattle, deforestation, soil erosion, and others. Yet even as these problems multiply, our society remains paralyzed. This paralysis stems from a conflict between our assumption that economic growth will continue and the environmental reality that we have already badly overshot any sustainable limits to growth.

Few people understand the reality of limits to growth on a finite planet. Even fewer understand the biological side of limits to growth: too many livestock, too many people, and the prospect of mass extinctions. Fewer still understand the social and economic implications of these limits; mainstream economists have completely dropped the ball.

If the economy has expanded beyond sustainable limits, then common sense suggests that we need a smaller economy. We must adopt three simple measures: (1) substantially reduce personal consumption, (2) substantially reduce human population, and (3) drastically reduce or eliminate livestock agriculture. This book explains why we need to do this and provides an overall guide how to do it.

First, a preliminary word of warning: while the measures I suggest are simple, implementing them is *not*. This book aspires to provide a radical and necessary *approach* to the environmental crisis. We understand quite a bit about what needs to happen and how it could be done, but we don't yet have all the details. Providing such details will require extensive collaboration among people with a variety of different perspectives.

You should not mistake this caveat, however, as doubt as to our ultimate destination. We can anticipate a world with much less energy; even renewable energy is limited by material shortages, the difficulty of supporting heavy industry, and energy storage issues. We can anticipate a world largely without livestock; livestock are now two-thirds of the megafauna biomass of the planet and have wiped out almost all wilderness and biodiversity. We can anticipate a world with substantially fewer people, because a truly sustainable economy cannot support eight billion humans, even if they were all vegans.

With luck, this new civilization may preserve many of the desirable facets of our civilization that are not dependent on destroying the biosphere and its flora and fauna. Perhaps this will include modern medicine, washing machines, agriculture, democracy, science, and books.

With a handful of exceptions, neither political groups, nor social justice movements, nor even environmental organizations grasp the magnitude of the problem. Those of us who *are* aware are almost alone in this struggle. Where shall we begin?

PART 1
THE PROBLEM OF LIMITS

CHAPTER ONE

IT'S NOT JUST CLIMATE CHANGE
The growing environmental crisis

Are there limits to the physical growth of the economy? Of course there are. We live on a finite planet with finite resources. These environmental limits, led by climate change, now threaten the basis of industrial civilization. At the same time, our society seems incapable of recognizing these challenges. These two realities—enormous problems and political paralysis—constitute what I call "the problem of limits."

It is common sense that the physical size of the economy depends on natural resources. There is only so much soil on the planet. There is only so much water, metallic ore, and fossil fuel. Most importantly, there is only so much atmosphere into which we can dump our various greenhouse gas emissions of carbon dioxide (CO_2), methane (CH_4), nitrous oxides, and all the rest. Since we live on a finite planet, the physical size of our economy must be limited as well.

With inventive technology and new discoveries, we can extend these limits considerably, but we cannot permanently avoid the issue of limits. We can improve the efficiency of our use of resources, but efficiency can never exceed 100 percent. We can make substitutions for minerals in short supply, but there are only ninety-two natural elements in the periodic table and sooner or later we are going to run through the ores we need. We can devise inventive technology, but sooner or later we are going to approach the limits of the possible, if we are not already there.

Reasons for Concern

Here's a quick overview of some current environmental challenges, which we'll explore in greater depth throughout the book:

1. *Climate change.* Each year brings disturbing news about the warming climate: melting Arctic ice and permafrost, a collapsing Antarctic ice shelf, ocean acidification, rising sea levels, greater storm intensity, giant methane sinkholes in Siberia, new and more devastating wildfires, and record temperatures. Average temperatures have already increased about 1 degree Celsius over pre-industrial levels, and we're approaching dangerous tipping points.

2. *Mass extinctions.* Species are going extinct at a faster rate than at any time since the extinction of the dinosaurs sixty-six million years ago when an asteroid hit Earth. Today, *we* are the asteroid: more than 95 percent of all the land-based megafauna (large animal) biomass of vertebrate species is now humans, their livestock, and their pets. Large *wild* mammals, including *all* the elephants, giraffes, zebras, bison, deer, bears, and apes, comprise less than 5 percent. Humans and their livestock have overrun the earth; there simply isn't that much space for wild animals. A sixth mass extinction, analogous to that which wiped out the dinosaurs, is already underway.

3. *Peak oil.* Numerous respected petroleum geologists have warned that there is only so much fossil carbon in the earth, and that we have used about half of all the economically

recoverable oil. Conventional oil production peaked in the United States in 1970; it peaked worldwide around 2008. Oil production has increased since then, but not by much and only due to increases from expensive and environmentally damaging "unconventional" oil (oil from fracking, deepwater drilling, etc.; see chapter 10). The extraction of unconventional oil may have bought us about a decade or two before oil extracted by *any* means goes into inevitable decline, with potentially civilization-ending results.

4. *Soil erosion*. Soil, the basis of our human food supply, is eroding at a rate ten to twenty times faster than the natural rate of soil formation in the United States and Europe. Soil erosion is even worse in Africa and Asia. Our agriculture is destroying the land that supports the expanding human population in the first place. So far, food production has been able to keep up with population increases, but obviously this isn't sustainable.

5. *Deforestation*. Forests have been progressively cleared for decades despite the protests of environmentalists. Cattle ranching is the main cause of the destruction of the Amazon rainforest, and recently political leaders have accelerated this destruction. Humans have destroyed about half of all the forest biomass in the world during the past ten thousand years. Besides being the home of countless millions of species, forests also contain the overwhelming majority of most of the earth's sequestered plant carbon, an amount far greater than all the carbon we've put into the atmosphere through burning fossil fuels.

6. *Water shortages*. Water is obviously indispensable for humans and all other species. There are no absolute shortages of water, but there is limited water availability in many agricultural areas. The Colorado River runs dry by the time it gets to the ocean, and this constitutes an absolute upper limit on water for the region without heroic and expensive water projects. Most groundwater-dependent agriculture, such as that in the Great

Plains and California's San Joaquin Valley, will eventually deplete that groundwater.

7. *World hunger.* With the depletion of agricultural resources, how long will we be able to feed even the current population of eight billion?[1] The Green Revolution—a set of new agricultural techniques utilizing high-yield crop varieties, irrigation, mechanization, and pesticides—has greatly increased the supply of food in the past fifty years. But the Green Revolution is also heavily resource intensive. It relies on inputs of depleting fossil carbon and declining water supplies.

8. *Population.* The exponential increase in human population in the past three centuries doesn't yet pose a fundamental limit on the economy—but only if we don't care about the environment or about human poverty. If the size of the economy could grow without limit, population wouldn't be an issue.

9. *Emerging infectious diseases.* Humans travel everywhere on the globe, crowd both domesticated and wild animals in close confinement, and give antibiotics to livestock with wild abandon. As a result of all these activities, animal diseases are now developing antibiotic resistance, mutating, and spreading to humans. Previously unknown emerging infectious diseases such as AIDS, SARS, MERS, and most spectacularly COVID-19, have all had a chance to spread in human populations. These have the potential, and sometimes the reality, of creating pandemics such as the 1918 "Spanish flu" and the current COVID-19 crisis, which have caused havoc among human economies and populations.

---

1   Eight billion as of November 15, 2022: United Nations Department of Economic and Social Affairs, Population Division, *World Population Prospects 2022.*

10. *Pollution.* Plastics, acid rain, heavy metals, traffic, industrial production, radioactive waste, and nitrogen fertilizers have all contaminated soil, air, rivers, streams, and oceans. Pollution has degraded not only the natural habitat of wildlife, but our human habitat as well.

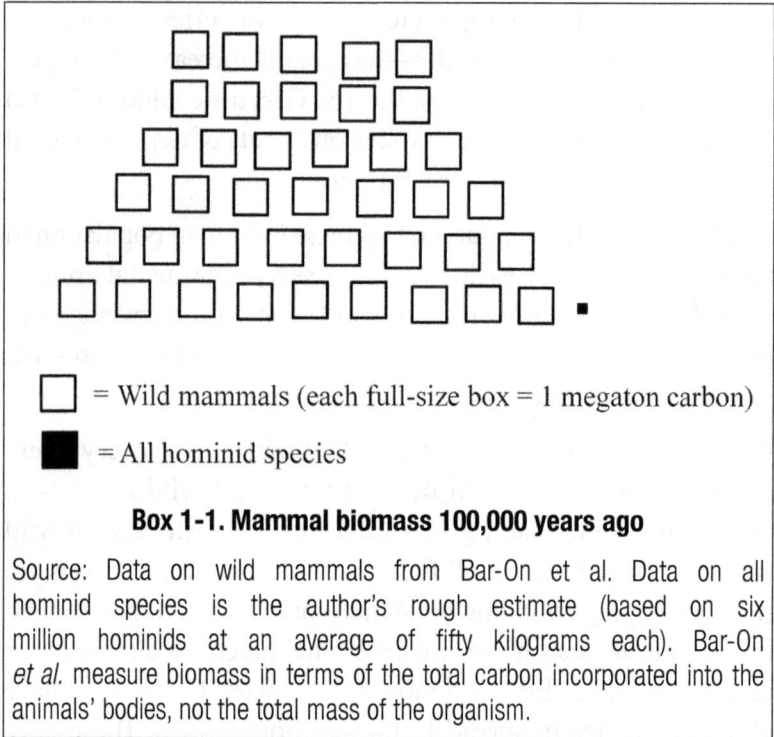

☐ = Wild mammals (each full-size box = 1 megaton carbon)

■ = All hominid species

Box 1-1. Mammal biomass 100,000 years ago

Source: Data on wild mammals from Bar-On et al. Data on all hominid species is the author's rough estimate (based on six million hominids at an average of fifty kilograms each). Bar-On *et al.* measure biomass in terms of the total carbon incorporated into the animals' bodies, not the total mass of the organism.

The complexity of dealing with limits to growth

The problem of environmental limits is a vast and sprawling topic, crossing many disciplines. These limits often interact with each other in complex and unexpected ways.

Climate change is perhaps the most important aspect of limits to growth. Yet our ability to cope with climate change is critically affected by *other* limits, as we will see later: limits such as the threat of peak oil, shortages of critical metals needed for energy-backup systems, and the destruction of forests. It's not that we can't deal with climate change. It's that we are likely to discover that there's much more to this task than building lots of wind turbines and solar panels. Fighting climate change will be messy and complex. We will need to make major sacrifices, which few are talking about and for which most of the world is totally unprepared.

A frequently overlooked aspect of limits to growth is the sheer mass of humans and their livestock, which overwhelmingly dominate all other large life forms on the planet. One hundred thousand years ago, humans were a small percentage of the land animals on the planet (Box 1-1, "Mammal biomass 100,000 years ago"). By 1900, humans were overwhelmingly the dominant species, and wild mammal biomass had declined precipitously (see Box 1-2, "Mammal biomass in 1900"). But just a century later, the biomass of humans and their livestock had increased explosively, while wild mammal biomass had declined even further (Box 1-3, "Mammal biomass in 2000"). Humans and their livestock today possess a carbon biomass *thirty-five times* more than all wild land mammals on the planet *combined*.

And humans don't comprise most of that biomass, livestock do! Other environmental complications go hand-in-hand with this near-total biological dominance: biodiversity collapse, extinctions, deforestation, soil erosion, and even climate change. Because this *biological* aspect of limits to growth is so consistently overlooked, and so obviously critical to our survival, I have made a special point of emphasizing it throughout this book.

Our crisis is much, *much* broader than any single issue, including climate change. These other environmental issues are just as serious as climate change, but still on the periphery of public consciousness. We don't face a single environmental problem but a *general* problem: the problem of limits to economic growth.

KEY:
☐ = Wild mammals
▨ = Livestock   (Each box is 1 megaton of carbon)
■ = Humans

Box 1-2. Mammal biomass in 1900

Source: data from Smil, *Harvesting the Biosphere* (2013), p. 228, table 12.2. Units are megatons of carbon. Smil also measures biomass in terms of the total *carbon* incorporated into the animals' bodies, not the total mass of the organism. Smil doesn't say, but we presume this is just the biomass of land-based mammals; Bar-On estimates that the biomass of fish today is several times that of all land-based mammals.

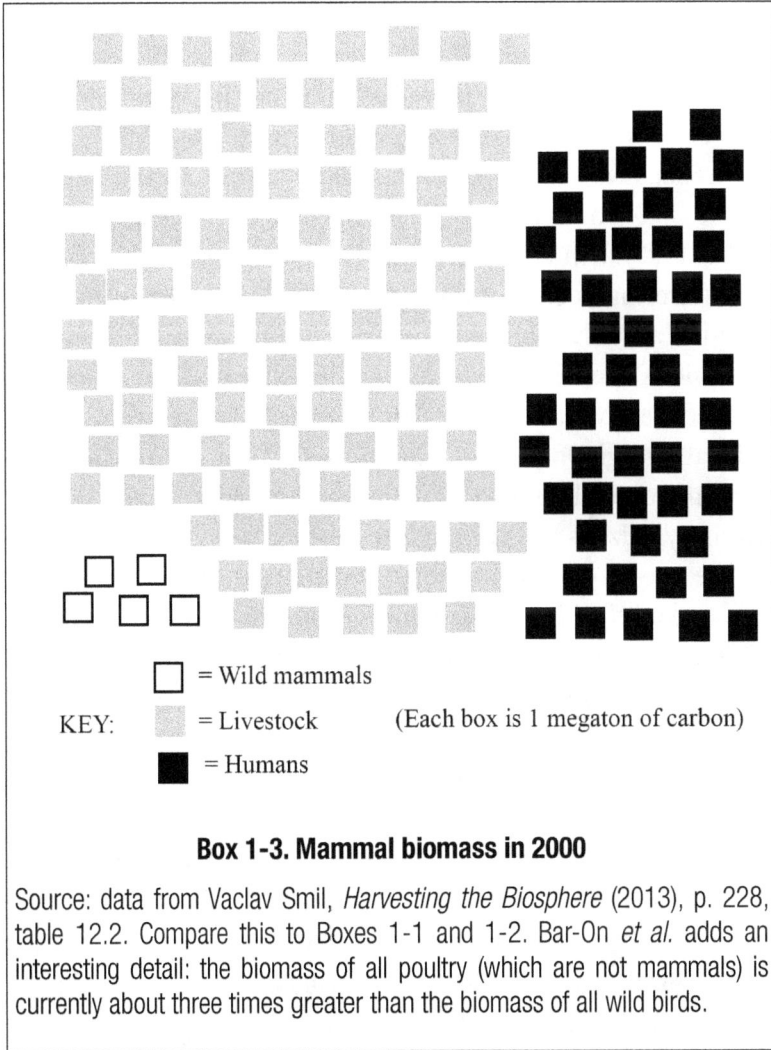

KEY:
☐ = Wild mammals
░ = Livestock   (Each box is 1 megaton of carbon)
■ = Humans

Box 1-3. Mammal biomass in 2000

Source: data from Vaclav Smil, *Harvesting the Biosphere* (2013), p. 228, table 12.2. Compare this to Boxes 1-1 and 1-2. Bar-On *et al.* adds an interesting detail: the biomass of all poultry (which are not mammals) is currently about three times greater than the biomass of all wild birds.

Conclusions

The days of easy solutions are long past. The 1970s featured the first "Earth Day" celebration; passage of the Clean Air Act and Endangered Species Act signed by a Republican president; a Democratic president advising us to adjust our thermostats to save energy; a World Vegetarian Congress in Maine; and the popularity of books such as *The Environmental Handbook, Diet for a Small Planet, The Population Bomb,* and *The Limits to Growth.* These books, flawed as they may be, have value and were certainly headed in the right direction.

Perhaps we should have also paid attention to the warning on the back cover of *The Environmental Handbook,* prepared for the first Earth Day more than fifty years ago[2]: "The crisis of the environment cannot wait another decade for answers." This sounded apocalyptic, and when succeeding decades produced no obvious apocalypse, we disregarded that advice.

And here we are. We are staring the collapse of industrial civilization right in the face. The subject of limits isn't going to be a popular one. Elites will resist change that reduces their wealth. Ordinary citizens will view radical proposals with apprehension.

We need a completely new and different economic system, modeled on the ideas of ecological economics, but we also need cultural changes and a different understanding of what it means to be human. We need a sense of shared vision in our societies that will undergird and make possible these changes. This is a tall order for today's fractured, polarized society, but as physical chemist and materials expert Ugo Bardi comments, "It is not impossible: in emergency situations, such as in wars or natural disasters, people understand the need of personal sacrifices for the sake of the common good."[3] In the coming chapters, I will show that *radical measures are necessary.*

....

2   De Bell, *The Environmental Handbook.*
3   Bardi, *The Limits to Growth Revisited,* 103.

CHAPTER TWO

UNDERSTANDING LIMITS
From empty world to full world

Our society is paralyzed from making any meaningful response to the environmental crisis. We struggle to reach a social consensus that climate change is even a real thing. We haven't even begun to seriously ponder peak oil, soil erosion, or anything else.

What's going on? Our difficulty is that we don't have a framework for understanding the problem of natural limits. Or do we?

The basic concept

Most economists disregard or dismiss the question of limits to economic growth. But a small minority of dissident "ecological economists" do provide a framework for such questions: that *we should view the economy as part of the larger environment.* Ecological economists frame the problem of limits as the contrast between an *empty world* and a *full world.* If the economy is part of the larger environment, then we need to ask, "How big can the economy get?" The world today is relatively "full" of humans, their animals, and their products—all of which dominate the planet.

The idea that the economy is part of the environment will strike most non-economists as undeniably true. Emphasizing that the human economy is part of a larger, physical world seems to belabor the obvious. But failure to understand this concept is a key reason why many mainstream economists and almost all political leaders haven't really paid attention to the problem of limits.

Early on in their standard textbook on *Ecological Economics*,[4] Herman Daly and Joshua Farley introduce the *empty world* and *full world* concepts (see Box 2-1, "From empty world to full world"). Ecological economists consider the *size* of the economy to be a *physical* concept, rather than a quantity measured by gross national product (GNP) or the money that trades hands.[5] Roughly, we live either in an "empty" world or a "full" world. What the world is "full" of (or "empty" of) is human beings and human-created infrastructure such as cities, roads, houses, cropland, livestock, computers, cars, food, and pollution. It can accommodate only so many physical things, depending on how clever (or lucky) humans are and how many resources the world has.

The terms "empty world" and "full world" are relative and conceptual ideas. Several thousand years ago, we lived in a relatively empty world, with smaller populations in rural areas and low use of energy and materials. It wasn't completely and literally empty; there were some humans and some resource consumption. But today, humans and their environments dominate the globe. Not only do humans dominate the world

---

4   H. Daly and Farley, *Ecological Economics*.

5   The size of the economy is the total of "throughput," or "the flow of natural resources from the environment, through the economy, and back to the environment as waste" (Daly and Farley, *Ecological Economics*, p. 6).

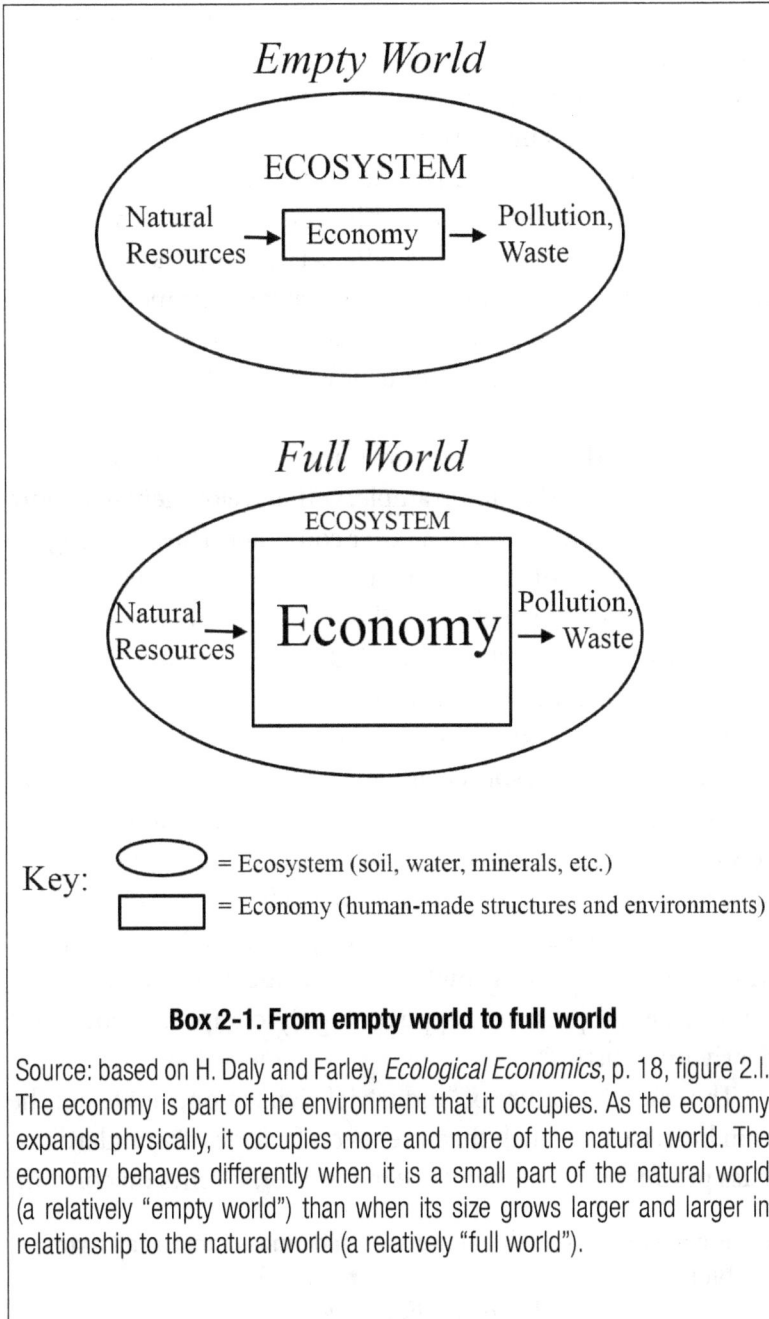

Empty World

ECOSYSTEM

Natural Resources → | Economy | → Pollution, Waste

Full World

ECOSYSTEM

Natural Resources → | Economy | → Pollution, Waste

Key: ⬭ = Ecosystem (soil, water, minerals, etc.)
▭ = Economy (human-made structures and environments)

Box 2-1. From empty world to full world

Source: based on H. Daly and Farley, *Ecological Economics*, p. 18, figure 2.l. The economy is part of the environment that it occupies. As the economy expands physically, it occupies more and more of the natural world. The economy behaves differently when it is a small part of the natural world (a relatively "empty world") than when its size grows larger and larger in relationship to the natural world (a relatively "full world").

of animals (Box 1-3), but we have also "filled up" the atmosphere with our greenhouse gas emissions. We no longer need to worry whether wild animals will eat us—rather, whether they will vanish from the planet entirely.

Again, the atmosphere isn't *literally* full; we could probably pump more greenhouse gas emissions into the atmosphere and manufacture more refrigerators. However, the economy's physical size has now reached truly alarming proportions; the total mass of all human-manufactured artifacts (including trash, concrete, and everything else) is now greater than the mass of all living things.[6]

However, the economy can improve or *develop* without increasing in total size: for example, with development of a more efficient irrigation system or a faster computer. Thus, "Ecological economics does not call for an end to economic development, merely to physical growth."[7] In theory, it is possible for human welfare to increase while the *physical* size of the economy remains constant or even declines.

This way of thinking is sometimes referred to as *decoupling* the economy from resource usage. So far, global GDP increases appear to correlate strongly with resource usage, and it's unclear how decoupling would work in practice. One group of European economists concludes that there is no evidence such decoupling is currently happening and that "the hypothesis that decoupling will allow economic growth to continue without a rise in environmental pressures appears highly compromised, if not clearly unrealistic."[8]

The essential nature of "industrial civilization," as we use the term here, isn't so much that it relies on factories or standardized mass production, but that it is *expansionist*. It relies on constant

....

6   Elhacham et al., "Global human-made mass exceeds all living biomass."

7   Daly and Farley, *Ecological Economics*, p. 64.

8   Parrique et al., "Decoupling debunked."

growth. Nothing about labor-saving devices or modern medicine, per se, *requires* continuing economic growth. But most of us can't even imagine a civilization without economic growth.

Malthus, Darwin, and exponential growth

Malthus was one of the first modern thinkers to appreciate the problem of limits. He first published his celebrated work *An Essay on the Principle of Population* in 1798, when human population was nearly one billion, compared to eight billion today. Malthus considered the relationship between unchecked population growth and the growth of food supply. By his calculus, if food supply increases over time "arithmetically" (e.g., 1, 2, 3, 4, 5, 6 ...), but population increases "geometrically" (exponentially, e.g., 1, 2, 4, 8, 16, 32 ...), then *population tends to outrun the food supply.*[9]

Strictly speaking, Malthus wasn't making a "limits-to-growth" argument as we understand it today, postulating an upper limit on human population at some figure such as nine billion or twelve billion. Malthus made no such calculations and foresaw that population *could* grow indefinitely, only with a limit on *rate* of growth: humans could increase in number, just not exponentially. But due to misery and vice, Malthus wrote, population growth is never unchecked. He was the first modern "prophet of doom"; without voluntary restraint, we would have war, famine, and disease to mitigate the excess population. Unfortunately, this worldview led Malthus to advocate cruel social policy. Better *not* to give charity to the poor, he argued: if you feed them, they'll just make more babies!

Today Malthus's conclusions are often held up for ridicule as exactly the reason we shouldn't accept a limits-to-growth argument. Since the publication of his essay, we have seen *both*

9   Malthus, *An Essay on the Principle of Population*, p. 4.

exponential growth in the population *and* exponential growth in the food supply. All we have known during the two centuries after Malthus has been what he seemed to think impossible.

But where did Malthus go wrong? While he is widely dismissed in terms of the human world, *his basic scientific insight is accepted without question in terms of the animal and plant worlds.* Charles Darwin saw that Malthus's general proposition for humans also applies to plants and animals. (Malthus, in a casual aside, noted that animals *also* reproduce faster than they obtain food—and so applied his "misery and vice" scenario to the animal world as well.[10]) For Darwin, the exponential reproduction of plants and animals implied an inevitable "struggle for existence" among them. In his autobiography, he recounts reading Malthus and immediately seeing the application to his theory of evolution:

> In October 1838, that is, fifteen months after I had begun my systematic enquiry, I happened to read for amusement Malthus on *Population* … it at once struck me that under these circumstances favorable variations would tend to be preserved, and unfavorable ones to be destroyed. The result would be the formation of a new species.[11]

Malthus's tenet drives Darwin's theory of evolution. Both thinkers agree that *all plants and all animals propagate themselves in excess of available food.* In *On the Origin of Species,* Darwin wrote, "A struggle for existence inevitably follows … It is the doctrine of Malthus, applied with manifold force to the whole animal and vegetable kingdoms."[12] The maximum reproduction rate ensures that as many members as possible of a species will

----

10 Malthus, *An Essay on the Principle of Population,* pp. 5, 8.
11 Nora Barlow, ed., *The Autobiography of Charles Darwin,* p. 120.
12 Darwin, *On the Origin of Species,* p. 66.

survive, but also that only the variations within a species most favorable for survival will be passed on.

"Misery and vice," for Malthus, are the natural checks which arise when population exceeds resources: things like hunger, disease, and fighting. If there were *always* plenty of resources, no matter how rapidly species reproduced, there would be no reward for favorable variations, no destruction of unfavorable variations, and no "struggle for existence." Species would never evolve but, we presume, just proliferate until they filled the whole universe. Therefore, it is these natural checks which arise when there is a shortage of resources—"misery and vice" in Malthus's language—which drives Darwin's theory.

Exponential growth in the twentieth century

Was Malthus wrong, or just ahead of his time?

The most obvious feature of the economy of the past five hundred years, compared to any previous period of human history and culminating in the twentieth century, has been its rapid and exponential growth, including population growth. Yet appreciating the enormous scale of growth in just the twentieth century is key to understanding *limits to growth*. Human population more than tripled in this period, but other indicators of physical growth increased even more. Water use increased nearly ninefold (Box 2-3), and energy use increased more than sixteenfold (Box 2-2). In the same time frame, we've also seen obvious advances in science and technology. But we are doing more than just using better tools. We are using and burning through resources at an unprecedented rate.

When political leaders and economists from both major political parties say they want to *continue* economic growth (or act as if this were inevitable anyway), we need to ask: *Really?* Do they believe that by the end of the twenty-first century, the economy will be ten times again larger than it was in the year

2000? Will this economic growth come from burning ten times the amount of coal, oil, and gas that we burned in the *entire* twentieth century? Or are we going to see a veritable orgy of wind turbine and solar panel construction, nuclear power, or what? And unless we start desalinating seawater, how will we expand water use *another* ninefold over the current already gargantuan use? Based on the accelerating destruction of the natural world for economic ends, we ought to give a different answer to these political leaders and economists: *this isn't going to happen.* The rate of growth will slow, and at some point, it will stop. The question is, when and how?

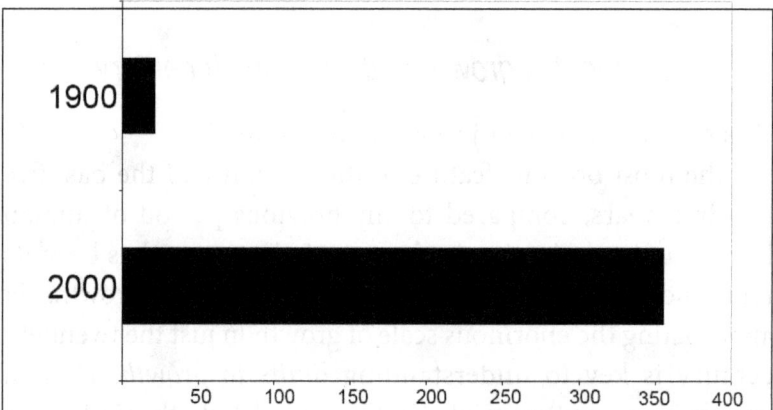

Box 2-2. World energy use in the twentieth century in exajoules

Source: data in Smil, "Energy in the Twentieth Century" (1 exajoule = 10^{18} joules). A joule is a unit of energy. An exajoule is more than 277 billion kilowatt hours.

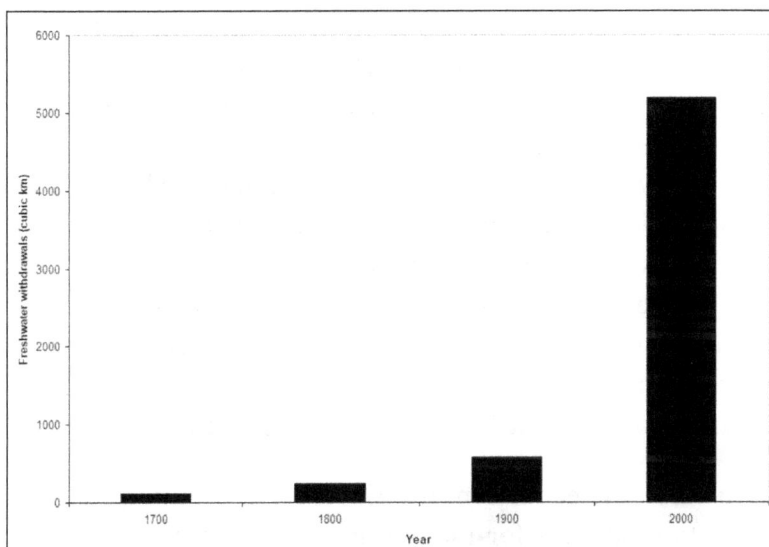

Box 2-3. World water use in the twentieth century (freshwater withdrawals in cubic kilometers)

Source: data in McNeill (2000), p. 121.

The problem of market failures

Ecological economists would likely agree with conventional economists on a vast array of issues including basic questions of supply and demand, which address the *allocation* of scarce resources. But there are two other aspects of economic policy that can't be reduced to questions of supply and demand. These are the questions of *scale*, or how big the economy can get, and *distribution*, or how we should distribute the economy's wealth. Scale is another way of talking about limits to growth or economy size, and distribution is another way of talking about inequality and social justice issues.

Economists often refer to the idea of a "free market," that is, economic exchange and trade in an unregulated (or relatively unregulated) trade environment. Buyers and sellers adjust what they buy and produce, and what they pay for these goods and services, based on their own economic self-interest. Celebrated eighteenth-century thinker Adam Smith argued that if everyone acts in their own self-interests, the aggregate result for the community is positive: sellers producing things that buyers want at a fair price, with competitive free trade determining what a fair price is and what buyers really want.

The free market works well for allocating resources in many cases, but does the free market address social justice issues or limits-to-growth issues? Consider a few notable examples of instances in which the free market was less than helpful:

1. In the nineteenth century, slavery was a quite profitable institution, at least for many slave owners. Could the "free market" have ended slavery? In principle, it eventually *might* have,[13] although this process evidently didn't work fast enough to prevent the American Civil War.

2. Today, someone might propose killing all the remaining elephants in the world, rendering them extinct, to sell the ivory. This could be a money-making proposition! Would a free-market system be able to prevent the extinction of the elephants? Again, in principle, the free market *might* eventually do this, because a human economy that allowed elephants to flourish might work better than an economy that wipes them out for their ivory or for their habitat. But so far, elephant populations are crashing, and the economy has seemed to reward anyone killing elephants or seizing their habitat.

3. What about climate change? James Hansen and other scientists have suggested that the threshold amount of CO_2 in the atmosphere, to avoid climate catastrophe, is no more than 350 parts per million (ppm). Yet today—with a vigorous free

---

13 The "profitability" of slavery has long been debated; for a survey of the arguments, see Aitken, ed., *Did Slavery Pay?*

market—CO_2 levels are over 400 ppm and rising steadily. The economy seems to reward greenhouse gas emissions, which continue and are *intensifying*.

In all these cases, if we trust a free-market approach, we *might* get lucky. It is possible that slavery in the American South would have disappeared anyway with the introduction of modern agricultural machinery, without the need for a civil war. It is possible that consumer boycotts or ecotourism would make it more profitable to keep elephants from going extinct. It is possible that solar and wind will become so cheap, they will entirely replace fossil fuels.

But the broader point is that *we shouldn't care* whether slavery, killing the elephants, or colossal greenhouse gas emissions is the best financial investment. Such choices are barbaric and stupid at the outset, no matter how much money can be made. *Values* are inherent in any discussion of sustainability or social justice—values that can't easily be reduced to *market* values. Doing nothing and letting the market take its course *already* involves us in value judgments.

Mainstream economists agree that there are such things as "negative externalities": unintended or uncompensated losses from an economic transaction, something that has negative consequences that wasn't part of the original transaction.[14] Slavery, species extinction, and climate change certainly qualify as negative externalities. The problem is that as we progress from a relatively empty world to an increasingly full world, these kinds of market failures seem to be growing in number and importance. Numerous resource problems are *not* adequately addressed by the prevailing economic system: overpopulation, species extinctions, soil erosion, groundwater depletion, and oil depletion.

We could make an analogy using the difference between

----

14 H. Daly and Farley, *Ecological Economics*, p. 175.

the Newtonian worldview and the Einsteinian worldview in physics. Newton's laws work quite well for a vast range of ordinary phenomena—they got humans to the moon, for example—but tend to break down if we look at the behavior of particles approaching the speed of light or the behavior of light approaching massive gravitational bodies. In the same way, the laws of conventional economics might work well enough in an "empty world," with fewer human structures but an abundance of diverse species, healthy soil, and water supply. However, they tend to break down as we approach a "full world" with plenty of human artifacts but less and less of the natural world. We can't ignore this problem anymore.

Conclusions

The framework of ecological economics offers an answer to the question of whether there are limits to economic growth. In broad terms, that answer is yes, there are limits to growth if we are approaching or living in a full world.

The paralysis of our economic system and our society generally, in the face of approaching limits, is palpable. What is causing our paralysis? Why doesn't everyone immediately embrace the ecological economics framework, accept limits to growth as self-evident, and start tackling some of these issues? It may seem like a simple proposition, but it's harder than it looks.

CHAPTER THREE

HARDER THAN IT LOOKS
Why dealing with limits is so difficult

Addressing the problem of limits to economic growth is an immense job. Why is this task so difficult? Some of the reasons are obvious. Others are less obvious, or not obvious at all, but still important. We need to keep these difficulties in mind as we investigate appropriate responses to environmental limits.

1. The general assumption of economic growth

Virtually everyone assumes that economic growth will continue. It's not hard to figure out why. Three centuries ago, industrial civilization began in England, and it subsequently spread around the globe. All we've known since is economic growth. There have been notable interruptions, such as the Great Depression of the 1930s, but each interruption has given way to the seemingly inexorable march of the economy upward with even greater force. Moreover, economic growth has been associated with vast improvements in the standard of living, ranging from labor-saving devices such as washing machines to advances in medicine such as anesthesia, vaccination, and birth control, not to mention computers and modern transportation. Meanwhile, predictions of doom due to insufficient resources haven't yet been realized.

2. The hostility of many elites

Every society has an elite class: a group of people—smaller or larger—who have a disproportionate amount of influence over political and economic outcomes. In a dictatorship or oligarchy, this class can be quite small. Our society is formally democratic but is no exception. In our society, elite class status is generally based on wealth, and such elites are typically able to manipulate the media, public opinion, and the political process to their liking. A small number of others, such as popular musicians, actors, preachers, sports figures, and so forth, may not begin by acquiring wealth but also break into this elite circle. None of these, of course, can *completely* ignore public opinion, which sometimes drifts from the path most elites would like it to take and which they sometimes learn to their discomfort. But the elites constitute the top 1 percent of the population and monopolize the direction of society.

The hostility of American elites to the concept of limits to growth may be, in the long run, the most significant problem in dealing with the problem of limits. If society is overconsuming, then we need to consume less. But the wealthy are the ones who consume the most and benefit the most from economic growth. Thus, the idea of limits to growth is an obvious threat to the American elite.

Those who benefit most from economic growth are unlikely to take kindly to suggestions that the economy needs to shrink. While American elites aren't completely united in this respect— there are wealthy progressives—the tendency of the elite class is to defend the interests of the wealthy, narrowly conceived.

To that end, they will use their wealth and political influence to shape political discussion and even scientific discussion. Special business interests, such as the fossil fuel, pharmaceutical, or livestock industries, will be especially eager to suppress any suggestion that their activities are harmful to the planet or to humans. These interests are profit-driven, and they will in turn use their influence to increase their profits—which, in a self-reinforcing pattern, in turn increases their ability to influence events further.

Truth and the accurate depiction of reality is an important tool. Ideally, we could turn to the scientific community for support, but this brings us to the next major obstacle.

3. The nature of science

In general, the scientific community is, or should be, our ally in ferreting out the truth and spreading awareness. One key problem—even since the time of Galileo—is outside pressure on science to distort its findings. Science needs financial support, which the wealthy will be reluctant to extend if it hurts their favored industries. The tobacco, energy, and livestock industries have all tried to create doubt about scientific studies that show the harmful effects of their businesses.

But there is another and more important problem. Even when science operates according to its ideals, it tends to be conservative, as Thomas Kuhn described in his classic book *The Structure of Scientific Revolutions*.[15] By "conservatism" we mean *scientific* conservatism, not political conservatism (scientists can be found all over the political map).

Scientific conservatism is simply the tendency of scientists to stick with what Kuhn calls "normal science." Scientists operate within an established framework of theories, ideas,

---

15 Kuhn, *The Structure of Scientific Revolutions*.

and assumptions (what Kuhn calls a "paradigm"). If evidence contradicts this paradigm, their first assumption is that the *evidence* is faulty, not the paradigm. Even if the evidence is valid, scientists will seek an explanation that leaves the basic paradigm intact.

Most of the time this is a good thing. A classic example from the history of science is the discovery of Neptune. Astronomers in the nineteenth century found that the path of the planet Uranus didn't conform to Newton's laws of physics. But they did not conclude that Newton's laws were false. Instead, they postulated a previously unknown body, which turned out years later to be the planet Neptune, and Newton's laws were confirmed again.

Scientists also tend to respect other scientists in disciplines outside their own. Physicists, though they may have private opinions, don't generally question basic conclusions of botany, economics, or cultural anthropology, and vice versa. They, like the public, are likely to defer to those who are generally accepted as the experts in their respective fields.

So far, so good: science tends to stick to its established theories; in itself, this isn't all that surprising and is in fact a good thing most of the time. But the conservatism of science can create problems, and we need to deal with at least three of them.

First, scientific conservatism complicates the relationship between science and politics. When political leaders ask scientists for advice about public policy, the scientific community typically strives for consensus that allows it to speak with a collective voice. This consensus implies that the science is sufficiently settled to be "decision ready"—to provide a basis for public policy. But to achieve community consensus, scientists tend to bias their voice in a direction that deviates the least from the default currently accepted view (Kuhn's "normal science"), resulting in a "lowest common denominator" effect. We see this in the scientific community's early estimates of the effects of climate change, where the "consensus" view, in retrospect, downplayed the seriousness of those effects.[16]

Scientific conservatism may also create problems if scientific issues don't clearly fall within one discipline or another. An urgent social problem may not clearly fall within any discipline, or it might legitimately fall within several disciplines—which is exactly the case with the problem of limits. The scope of the topic *limits to growth* is *colossal*. Looking at the subjects of population, peak oil, climate change, land use, world hunger, economic collapse, consumption patterns, mass extinctions, and social change relies on informed judgments in the fields of physics, chemistry, biology, climate science, nutrition, petroleum geology, psychology, sociology, history, paleontology, agriculture, ecology, and economics.

Finally, scientific conservatism can be a problem if a discipline that affects a matter of vital public interest becomes stuck within a flawed paradigm. This is exactly the situation with most economists' narrow ideas about economic growth.

Scientific revolutions aren't frequent, but they do occur. Scientific disciplines will discard their basic theories and reach a new understanding and a new paradigm. But this isn't an easy or a quick task. This is what happened in the transition of physics from the Newtonian worldview to the theories of Einstein. The transition from Newton to Einstein had two earth-shaking implications: the discovery of nuclear energy and the invention of the atomic bomb. It took a long time for the implications of this scientific revolution to unfold; four decades separated Einstein's special theory of relativity from the detonation of the first atomic bomb.

What should we do about the flawed paradigm of economics? If we lived in an ideal world, we could simply wait for this scientific revolution in economics to play out. However,

---

16 Oppenheimer, M. et al., pp. 10–17.

economics today is an area of urgent public policy, and we likely don't have four decades for economists to work these problems out on their own. We need to be objective, but in a hurry. That brings us to our next difficulty in dealing with limits to growth.

4. The crisis of economics

When we look around for allies in dealing with limits to growth, economists would seem to be a logical choice. Aren't they the experts on "economic growth"? Unfortunately, economists are generally dismissive or unhelpful on this problem, and many of them *fight* the idea of limits. Economics is now going through a crisis as it seeks to understand a world that it is increasingly failing to describe. In the meantime, political leaders and the public are relying on "experts" who are operating from within a flawed paradigm.

If there was ever a chance for mainstream economics to address the problem of limits, it was after *The Limits to Growth* was published in 1972. This book put forward "world modeling" ideas, based on early computer programs, that projected resource scarcity by the middle of the twenty-first century. There were plausible arguments against these projections; for example, some argued that technological advances could offset resource decline. But proponents of the world-modeling ideas in *The Limits to Growth* were ostracized and encountered little except prejudice in the scientific world. Ugo Bardi notes that "the debate about world modeling was especially harsh and ... did not follow the accepted rules for this kind of exchange."[17]

The popular press followed the lead of the economists. In a 1989 article titled "Dr. Doom," popular author and journalist Ronald Bailey claimed that according to predictions in *The Limits to Growth*, we should have run out of gold, mercury,

....

17 Bardi, *The Limits to Growth Revisited*, p. 59.

and tin already and would run out of petroleum, copper, and several other minerals by century's end. But in fact, *The Limits to Growth* said nothing of the kind. Bailey's article was widely quoted throughout both the popular and the scientific literature, without anyone bothering to check to see if his claim was true.[18]

One of the early critics of *The Limits to Growth* was William Nordhaus, who subsequently won the 2018 Nobel Prize in economics. In his book on climate change, *The Climate Casino*, Nordhaus predicts fantastic growth in the economy for at least another century or two. According to Nordhaus, the *average* world per-capita income (in inflation-adjusted dollars) will be $55,000 in 2100 and $130,000 in 2200![19] By contrast, today it is less than $10,000. Nordhaus then maps out a climate strategy whereby this fabulous future economic growth will help pay for measures to address climate change.

He says, somewhat apologetically, "This sounds like a fantasy, but it is the result of exponential growth of living standards."[20] It is as if he intuitively understands the absurdity of assuming indefinite exponential growth but is being forced by his mathematical extrapolations to do so. This doesn't *necessarily* mean that Nordhaus is wrong, but perhaps it sounds like a fantasy because it *is* a fantasy.

Economic assumptions are now being challenged, not just by ecological economists, but also by economics students themselves. Young rebels have written open letters to professors, organized themselves into a global network, and disrupted economic conferences. "No other academic discipline has managed to provoke its own students—the very people who have chosen to dedicate years of their life to studying its theories—into worldwide revolt."[21]

18 Discussed in Bardi, *The Limits to Growth Revisited*, pp. 88–91.
19 Nordhaus, *The Climate Casino*, chart on p. 80.
20 Nordhaus, *The Climate Casino*, p. 81.
21 Raworth, *Doughnut Economics*, p. 3.

5. The problem of collapse

The failure of mainstream economists to acknowledge the problem of limits makes it much more likely that there will be an unexpected economic and political collapse of some sort. Political leaders, understandably enough, typically rely on economic "experts" for judgments about economic matters. If the economists don't know what to do, it is unlikely that political leaders will know what to do, either. We don't know exactly what the consequences will be if we fail to deal with these various environmental problems, but it *will* have consequences, and not just for the animals and the earth. It will have consequences for *us*.

Where will that leave us? We don't know. Collapse is a *political* event,[22] not a mechanical result of resource shortages, as we will see when we discuss this important and complex topic in chapter 12. Possibly, a collapse would lay out the consequences of economic growth for all to see, and we could start building a new world on the ruins of the old. Or it might result in a descent into a barbaric world with a more primitive technology and social order.

For now, what we need to know about collapse is that while it doesn't necessarily mean the "end of the world," it makes planning difficult. A slight variation in events might produce quite different outcomes, and consequently, we don't know what to plan for. We can make out the probable initial impact, but not our own human response. We should prepare to deal with the consequences of inaction; we are already seeing some of them.

6. The problem of collective action

I often use the term *we* in this book, as when I say things like "we live on a finite planet with finite resources." The exact

----

22 Tainter, *The Collapse of Complex Societies*, p. 4.

meaning should be (I hope) clear from the context. Usually it means something like "the nation, society, or all the peoples of the world." It can sometimes mean "the author and readers of this book," as when I wrote earlier, "We need to keep these difficulties in mind as we investigate appropriate responses to environmental limits."

There is a third, more ambiguous meaning of this *we*: when we ask the question, "What do we do now?" The question "What do we do now?" seems to make perfect sense when we say it out loud, and collective action is obviously more effective than the actions of a single individual acting alone. But who is this "we"?

"We" in this case refers roughly to the group of people who are aware of the problem of limits and are willing to take collective action. But not everyone who is *aware* of the problem is going to be aware of, or able to coordinate with, others who are aware—and their awareness may vary quite a bit. Even people who are aware will have varying viewpoints on what should or could be done. And this "we" is going to vary from time to time. At the present time, the number of people who are aware of the problem of limits is vanishingly small, though it appears to be growing. Some of the people who are aware are "doomers"—those who feel that civilization is doomed no matter what we do—and thus may not feel inclined to do anything at all. Others may have already decided on their own particular set of solutions—perhaps focused on renewables, birth control, or veganism.

Any literal answer, therefore, depends not only on our proposals to change government policy, individual behavior, or group behavior, but also on whom we can mobilize in support of whatever we are proposing. People naturally don't want to be left with a book that gives them a long description of various environmental outrages, but no suggested plan of action, or (what is almost worse) just a few platitudes in the final chapter.

We can define the necessary *approach* to this problem, but we cannot spell out all the details. Just as the problem of limits is sprawling and immense, so the possible courses of action are sprawling and immense. Who "we" are, when we try to identify "what we can we do", therefore, will need to remain ambiguous for the time being, but we (the author and readers of this book) should be thinking about this as we go along, and we will return to this question at the end.

Conclusions

Almost everyone assumes that the economy will and should continue to grow. The scope of the task of dealing with limits is enormously complex anyway. The economists, who should be our chief allies, are largely dismissive or even hostile to the whole enterprise. The economic elites in our society, well educated but too often self-absorbed, are typically hostile to placing any limits on economic growth. The scientific community is, by its nature, conservative on scientific questions. Uncertainty about the future, and uncertainty about precisely whom we can mobilize behind any proposed actions, adds another layer of difficulty to our task. Since inaction on the environment is likely, catastrophe is *also* likely.

Such a catastrophe could take a variety of different forms. A decline in resource availability might lead to financial collapse, like the Great Depression of the 1930s, or the one that almost happened in the wake of the 2008 financial crisis, or the one that may still happen as a result of the COVID-19 pandemic, the Russian invasion of Ukraine, or some fresh outrage. Economic problems also might lead to political instability that generates demagogic leaders, international conflict, or civil war. It might be a quick collapse, or it might be a slow, gradual slide. The unpredictability of both such a collapse and its consequences then becomes another aspect of the problem.

It isn't possible within the scope of this book to include a full discussion of everything relevant to the topic of embracing limits. What we seek to do here is provide a broad overview of problems which aren't widely discussed but are critically important to the survival of human civilization. Our aim is to create a level of shared vision that makes it clear that *radical measures are now necessary*.

CHAPTER FOUR

DEGROWTH

Getting to a smaller economy

Environmental limits to economic growth involve technical issues, but the problem of limits isn't primarily a technical problem. It's a *social* problem and requires social changes. We have enumerated key obstacles we face in dealing with this issue; now let's look at a broad outline of how to approach it.

If the economy is too big in relationship to resource limits, then the answer must be a smaller economy. This should be common sense. And if our ultimate goal is to get to a smaller economy, then we need the opposite of economic growth—namely *degrowth*.[23]

In looking at individual limits (such as energy, soil, biodiversity, etc.), we will ultimately want to know several things. First, how do these individual limits relate to the *overall* problem of limits? Second, how does the problem of limits affect

---

23 "Degrowth" as a topic is almost completely unknown in the United States, but is more actively discussed in Europe. A good introduction to this discussion is Barlow, Nathan, et al., *Degrowth and Strategy*; a more popular treatment is Hickel, *Less is More.*

questions of economic and social justice? Third, how does the problem of limits affect our relationship to the non-human world?

1. Individual limits versus overall limits

By all means, we should join with movements (such as the climate movement) to preserve the environment. However, when confronting individual crises, we need to understand the *overall* problem of limits, not just that one particular limit.

Nowhere is this more apparent than with climate change (discussed more fully in chapter 14). While the reality of climate change is well discussed, what to do about it is *not*. In the climate movement, the political consensus seems to be that we should replace fossil fuels with renewable energy: wind turbines, solar panels, biodiesel, and hydroelectric dams. With renewable energy, the thinking goes, we'll be able to do all the same things we're doing now but do all of them sustainably. Proponents of the "Green New Deal" in the United States usually don't put this in so many words, but it's the unstated conclusion they seem to encourage people to draw.

Renewable energy is unlikely to work like this. Even if renewables work like a charm (and it's unclear that they will), they *still* require resources. Wind turbines to power the entire world's current energy consumption would require enormous quantities of steel and concrete, much more than that required to deliver an equivalent amount of energy from coal or oil. They also require physical space; the number of suitably windy sites in the United States for wind turbines is large, but not infinite. And they do nothing for soil erosion, groundwater depletion, deforestation, or species extinction.

None of this means we shouldn't build renewables anyway! But renewable energy won't solve the problem of limits to growth. At best, it deals with one particular limit—namely, climate. And dealing with climate will require economic sacrifices from the American public, for which the public is almost totally unprepared.

Similar issues continue to plague the Green Revolution (discussed more fully in the next chapter). The Green Revolution, instituted in the 1970s, substantially increased crop yields and likely averted disastrous famines at the time. It did so through use of irrigation, fertilizers, pesticides, and new high-yielding crop varieties. But it didn't permanently solve the problem of world hunger. Rather, instead of being limited by the total amount of land, food production is now limited by the total amounts of water and energy. And because population has now doubled *again* since the Green Revolution started, we are likely to face, again, the same issues that the Green Revolution addressed in the first place. We solved one problem but now must confront others.

Both these examples illustrate "Liebig's law of the minimum," popularized by Justus von Liebig, a nineteenth-century German chemist. When multiple inputs are required for any process, the resource that is relatively the scarcest (not necessarily the resource that is scarcest in absolute terms) is the limiting factor for the overall process. Though this principle carries the title of a *law*, it is common sense for any cook. If you have a limited amount of yeast, there is only so much bread you can make, no matter how much flour and water you have.

We confront *multiple* limits, and we need to look at the overall problem of limits to growth, not just this limit or that limit. A piecemeal approach to the environmental crisis won't work; in fact, it could make things even worse by creating the illusion of progress when we are standing still or even moving backward. At best we have shifted things around from one limited resource to another limited resource that is just a bit more abundant. The only real and permanent solution to the problem of limits is the sustainable use of renewable resources *and* a smaller economy.

2. An imperative for social justice

Talking about degrowth or a smaller economy often evokes scary images of economic recession, depression, or worse, which we need to address. If we now decrease the size of the economy, won't there be less to go around, and thus a decline in the standard of living? That surely won't go over well with the public.

Asking people to get along with less is hard enough even when an acknowledged crisis looms. It is unlikely to work at all as long as we have vast and increasing wealth inequality. It can only work if it's crystal clear that we're all in this together. We have no real experience with economic declines, except for recessions or wartime conditions understood as merely temporary. All our society has known in the past few hundred years, at least in the West, has been a general increase in the standard of living.

We want to do more than just formulate "austerity" measures. Austerity in the form of across-the-board cuts in consumption is obviously unfair. For the wealthy, such cutbacks would be just an inconvenience. But for the poor, a proportionally similar cutback would be devastating. The poor and the homeless don't have anything to cut back on! The "yellow vest" protests in France in 2018, initiated in response fuel tax increases, demonstrated that such measures tend to increase instability. The burden of implementing degrowth must fall on the wealthy, whose outsized consumption is responsible for most of the environmental destruction in the first place.

This means that we need to examine social justice issues and redress economic inequality in tandem with retooling the economy to protect the environment. Social justice is an environmental issue; without social justice, degrowth is likely to be quite unpopular. The flip side to this is that resource limits on the economy mean sooner or later the economy *will* contract; Mother Nature will enforce her own degrowth policy. At that point, economic growth—creating more wealth as a method for alleviating poverty—won't work at all. This will push demands for social justice in a more radical direction, namely the direction of income redistribution (discussed further in chapter 21).

3. A cultural and ethical dimension

There is something strangely perverse about needing to point out that our poisoning of the atmosphere, our ripping apart of the biosphere that supports photosynthesis, and our driving wild animals to extinction—not to mention the torture and slaughter of billions of domesticated animals—has disadvantages to human beings as well. It's as if without the consequences for humans, these destructive behaviors would be perfectly acceptable.

Do we need to say that these outrages that have wrought great suffering should be off the table from the outset? It's our *destruction* of the earth that should require careful justification, not proposed attempts to undo and reverse it.

There is a cultural dimension to embracing limits. There are *compassionate* and *ethical* dimensions to embracing limits. Somehow, we need to develop and expand this environmental consciousness until it becomes second nature to human beings, so that we don't have to refight this struggle over and over.

To get to that point requires more than individuals waking up and reaching such conclusions. We need a *collective* awakening.[24] Such a transformation may seem tremendously ambitious, but at some level and in some way, it has happened before. Each

24 As Thich Nhat Hanh says, "In order to save our planet Earth, we must have a collective awakening. Individual awakening is not enough." Nhat Hanh, *One Buddha is Not Enough*, vi.

generation doesn't need to debate whether crimes against other humans, such as wanton killing, lies, or stealing, are wrong. For almost everyone (except the mentally ill), this goes without question. Thus, stating the truth in and of itself becomes a force for good, as Gandhi illustrated in his *satyagraha* (truth-force) campaigns. In a time of social decline, of course, these mores are subject to abuse, as corruption flows throughout the social order. But even in our corrupt order, there is a yearning for a return to a simpler time when *yes* means yes and *no* means no.

Many people are aware that all the great religions, as well as secular systems and philosophies, tend to converge in their practical ethics, at least toward humans. The principle of "do unto others as you would have them do unto you," is found, in different forms, in all the major religions. This "Golden Rule" is so ubiquitous that some might assume it is universal to all societies, but it is not. Before the Axial Age (c. 800 BCE to 200 BCE), morality was largely limited to the tribe. While there were occasional universalist ethical thinkers before this time (Hammurabi, Moses), this kind of universal morality came to the fore during the Axial Age and was associated with the development of great empires such as those in ancient China, Greece, India, and Rome. If everyone feels bound *not* to kill, lie to, or steal from others (even those outside of one's immediate tribe), it facilitates relative peace and harmony in society and the world. As we will discuss later, this moral sense is behind the evolution of these empires.

Today, we need a similar development in universal morality and culture, but instead of mandating moral treatment only of other human beings, it must mandate moral treatment of all life on the planet—plants and animals as well as humans. A key reason we are in an environmental crisis is that our current ethical understandings are inadequate. Many people accept the importance of treating other *humans* ethically. We need to have respect or reverence for the lives not only of humans, but also for all plants, animals, ecosystems, and even minerals on the planet. It is not only wrong, but highly unwise, to behave otherwise.

Conclusions

Degrowth, as we talk about it here, implies more than just an economic program. We need to approach the scientific questions involved in limits to growth by understanding the interdependence of many different factors. We need to be aware that the problem of limits has social consequences as well as economic impacts, and constantly keep in mind the importance of social justice—without which our task will most likely be impossible. Finally, there is a spiritual dimension to embracing limits, which needs to be embedded in human consciousness just as the universalist ethics of the Axial Age thinkers are embedded in today's cultures and religions.

The problem of limits has only one real solution. That solution is a smaller economy—a substantially smaller population, substantially less consumption, and drastically reduced or eliminated livestock agriculture—with a society, ethics, and culture to match.

PART 2
THE STATE OF THE PLANET

CHAPTER FIVE

BIOLOGY FOR A SMALL PLANET

Plants, animals, and humans living together

Many people today are rightly concerned about human overpopulation. Climate change, biodiversity collapse, and peak oil are all problems caused by humans that also negatively impact humans. All these problems and their solutions are going to scale up with human population.

We certainly need to ask how many people the earth can support (see chapter 18). But this begs an even more critical question: Does that leave room for anything else on the planet? If we are determined to use every last square inch of the planet for humans, what will become of all the plants, animals, and the rest of the nonhuman world? And won't how we treat *them* ultimately rebound on *us*?

Understanding human impact

How many people can the earth support? Obviously, it depends on how consumptive these humans are. We don't really have a human population problem; we have a human *impact* problem. Affluence typically increases per-capita impact, though technology (at least potentially) decreases per-capita impact. We need to consider these factors as well as absolute population numbers. This approach is captured in the classic "IPAT" population-impact equation developed in the 1970s by Paul Ehrlich and John Holdren:[25]

$$\text{Impact} = \text{Population} * \text{Affluence} * \text{Technology}$$

However, while the IPAT equation isn't exactly wrong, it can be misleading, depending on the context in which it is invoked. The IPAT equation leaves us with the vague feeling that we can accommodate any arbitrary number of people on the planet by adjusting affluence or technology up or down, allowing us to evade the question "How many people can the earth support?"

To a certain extent, we *can* accommodate more people on the planet through technology. There is one spectacular example from the 1970s in which technology *did* get us out of a serious jam: the Green Revolution (discussed later in this chapter). The Green Revolution dramatically increased agricultural yields and probably averted serious famines throughout the world. This was great news but also, unfortunately, led many people to assume (mistakenly) that we no longer had to worry about feeding an expanding population—that we could just scale up technology to achieve any kind of result.

Nor can we arbitrarily reduce affluence *downward* indefinitely to accommodate increasing numbers of people. There is a minimal level of "affluence" needed just to stay alive. We can eat beans and rice, but we have to eat *something*. This minimal level of affluence is already the lot of many of the world's poorest people, as *billions* live on just one or two dollars a day—less than what it costs to care for a typical American dog or cat.

....
25 Holdren, "A Brief History of 'IPAT.'"

Moreover, inequality has a highly destabilizing effect on society—as our highly unequal society today is now discovering. Resource limits already constrain attempts to give those at the bottom of the social order a decent standard of living. It is only in the past few decades that China and India have begun to do this; they now chafe at demands from other countries that they limit their coal consumption due to climate concerns.

The population-livestock problem

There's a second and even more serious difficulty with the IPAT equation. By focusing on human impact, it obscures the impact of livestock. A key driver of environmental limits is the sheer mass of humans and their livestock, which overwhelmingly dominate all other large life forms on the planet. The biomass of the global livestock population is twice as large as that of the human population!

The increase in human population from even a few centuries ago is breathtaking enough. Human population has increased one hundredfold since the time of the ancient Pharaohs in Egypt, and more than tenfold since the beginning of the Industrial Revolution.

But in the meantime, livestock population has increased even *more*. The total carbon biomass of *all* mammals in prehistoric times, 100,000 years ago, was about 40 megatons, which dwarfed that of all hominid species[26] (see Box 1-1). By the year 1900, total mammalian biomass had increased substantially (Box 1-2), and by the year 2000, it had exploded (Box 1-3) fourfold.

But this was good news for only one species: us. One hundred thousand years ago, when all mammals eclipsed the insignificant population of all hominid species, there were also

....

26 Besides *Homo sapiens*, there was *Homo neanderthalensis*, *Homo erectus*, and *Homo floresiensis*. These other three, however, had disappeared by ten thousand years ago.

no "livestock" to speak of. By 1900, the total carbon biomass of humans was roughly equal to that of wild mammals. By 2000, however, humans and their livestock possessed a carbon biomass *thirty-five times* greater than all wild land mammals on the planet *combined*. And most of that biomass isn't humans— it's our livestock. In the meantime, the broad diversity of *wild* mammals 100,000 years ago has diminished, with many species going extinct and many others in danger of extinction.

Biologist and ecologist Anthony Barnosky presented data that both corroborates and contrasts with the data cited in Boxes 1-1, 1-2, and 1-3. Barnosky estimates that 100,000 years ago, the total megafauna biomass (humans, domesticated animals, and large wild animals combined) was about 200 billion kilograms, whereas today it is 1,500 billion kilograms.[27] That makes the increase in biomass even greater: more than sevenfold.

We are, to a certain extent, comparing apples to oranges here. Smil (and Bar-On et al.) are talking about total *carbon* biomass[28] and are talking about *all mammals*—presumably including smaller mammals such as squirrels. Barnosky, by contrast, is talking about *total* biomass (which includes the weight of water and other materials as well as carbon) of all *megafauna* (large animals greater than one hundred pounds in weight). Despite the differing approaches and differing numbers, these three estimates still agree on the essentials: (1) mammalian biomass has exploded in the past 100,000 years, (2) livestock and humans dominate large animals on the planet, and (3) wild mammals are in steep decline. The sheer weight of livestock is more than *double* the weight of all humans, and it is also increasing *faster* than human biomass.

....

27 Barnosky, Anthony D. "Megafauna biomass tradeoff as a driver of Quaternary and future extinctions." The Quaternary is the most recent geological period, starting about 2.6 million years ago.

28 Humans have about 18 percent of their total biomass as carbon; you can't compare Smil's figures with Barnosky's directly.

We have heard about the "population bomb," but what about the "livestock bomb"? We should speak not of the "population problem" but of the *"population-livestock* problem" because many of the same general resource constraints limit both human population and livestock population. Obviously, humans have more impact on the resources needed to supply consumer goods, such as cars, flat-screen TVs, and trips to exotic destinations. But in terms of *food* resources (land and water), the requirements of humans and livestock are roughly comparable.

Some animals can digest foods that humans can't, such as cows eating grass, so cows don't necessarily compete for *human* sources of food. But as we will see later, the cattle industry is extremely destructive of other resources, such as forest land and wilderness area. From the point of view of the biosphere, cows who "need" a forest converted to pastureland aren't much different from humans who need to convert a forest to cropland. Technology can help here; we can develop more efficient irrigation systems, create new more productive plant varieties, and breed animals to gain weight rapidly (cruel, but efficient). But there are certain minimal allocations in terms of soil and water that we can't get around.

In fact, the environmental impact of livestock is *greater* than the environmental impact of the same biomass of humans. Food animals are typically all slaughtered as youngsters—the equivalent of human teenagers. As soon as they grow up, they stop growing. There is no economic return in continuing to feed them, so they are slaughtered. Dairy cows and egg-laying hens last a bit longer, but still die young. A growing animal takes much more food (per unit weight) than a comparable adult animal. Adult animals just need food to maintain their weight; growing animals need to maintain their weight *and* increase it.

The biomass question

What we really face is not limits on human population, but limits on *biomass* in our ecological niche. Our ecological niche is that of a large animal—one of the "megafauna," typically defined as those animals who (as adults) weigh at least one hundred pounds. Because of our tools, weapons, and agriculture, we have both increased agricultural productivity *and* expanded the territory we control (which, at this point, is essentially the entire land area of the planet).

We have no real ecosystem enemies among the other megafauna. All we have to worry about is resources. By wiping out the rest of the wild megafauna—by killing off all the elephants, giraffes, bison, and so forth—we wouldn't increase our access to land resources or our safety against predators by all that much. The real question is not how many *people* the earth can support, but *how much megafauna biomass the earth can support.*

It would make sense to say that there is a "natural" maximum extent of megafauna biomass of about 200 billion kilograms, or 40 billion kilograms of carbon biomass (= 40 megatons). Barnosky clearly suggests something like this, without committing himself to a particular megafauna biomass number, in his idea of a "biomass tradeoff."[29] That is, in natural conditions, with limited land area, there will be an upper limit on plant photosynthesis and thus roughly some maximum extent of plant life. Since animals in this ecological niche depend on plants (directly or indirectly) for food, there is a maximum extent of such animal life as well. To push beyond this "natural"

....

29 "As human biomass grows, the amount of solar energy and net primary productivity (NPP) available for use by other species shrinks, ultimately shrinking the amount of the world's biomass accounted for by those non-human species." Barnosky, "Megafauna biomass tradeoff," p. 11543.

level requires something like agricultural technology.

Barnosky estimates that by ten thousand years ago, overhunting by humans had *reduced* megafauna biomass by about half. From ten thousand years ago until about the year 1500, humans and their domestic animals slowly increased to fill this megafauna gap, so that by 1500 there were once again roughly 200 billion kilograms of megafauna biomass. During this period there were relatively few species extinctions of wild megafauna. The African island nation of Madagascar, where humans didn't enter until about two thousand years ago (much later than they entered the major continents), is one of a handful of exceptions that prove the rule. A major extinction episode in Madagascar followed shortly after humans discovered this large island.

But after 1500, this all changed. Total megafauna biomass went from 200 billion to 1,500 billion kilograms, *more than seven times greater*, in what is a geological blink of an eye. Humans comprised about one-quarter of this increase, their livestock about three-quarters.[30] Wild animals declined both in absolute and relative terms, swamped by the increase in humans and their livestock.

There are more total megafauna on the planet than there have been for at least 100,000 years. There is now more *livestock* biomass on the planet than there has been of *all other large animals (including humans) combined* at any time during the past 100,000 years. Meanwhile, biodiversity is crashing as wild animals disappear and species go extinct, a consequence of the human domination of the biosphere, a process which dramatically accelerated just in the twentieth century (once again: see Boxes 1-1, 1-2, and 1-3).

....

30 Barnosky, *Dodging Extinction*, 58.

Year	Cropland	Grazing land
1700	3 million km^2 (1) — 6 million km^2 (3)	5.3 million km^2 (2)
1860	5.7 x 10 million km^2 (5)	
1900		13 million km^2 (6)
1920	10.0 x 10 million km^2 (5)	
1978	14.3 x 10 million km^2 (5)	
2000	15 million km^2 (4)	33 million km^2 (2)

Box 5-1: World agricultural land use, 1700–2000

Sources:

1. Matson et al., "Agricultural Intensification and Ecosystem Properties." Matson et al. implies that it has increased 466 percent, therefore must have been ~3.19 million square kilometers (km^2) in 1700, and cites Meyer and Turner, "Human Population Growth."

2. Asner et al., "Grazing systems, ecosystem responses, and global change." For the year 2000, I prefer this source to Ramankutty et al., in "Farming the Planet"; Asner et al. includes arid lands and grazed forest lands (p. 17), important when considering habitat destruction.

3. Ramankutty and Foley, "Estimating historical changes in global land cover."

4. Ramankutty et al., "Farming the Planet." This figure disagrees with the figure given for cropland (18 million km^2) in Ramankutty and Foley, but this is due to a change in methodology.

5. Williams, "Forests," p. 183.

6. Smil, *Should We Eat Meat?*, p. 151.

And what about plants?

Humans dominate the world of large animals on the planet, but they also dominate the *plant* world. In the world of plants, as well as the world of animals, humans are limited simply because there's not much left to exploit that we haven't already exploited. Humans have managed to break through many natural barriers—first with the invention of agriculture, then extended further into the twentieth century by the Green Revolution. But while human ingenuity has stretched natural limits considerably, we cannot permanently evade them.

We should note the impact of humans on the plant world from the vantage points of agricultural land use (it's immense), total plant biomass or *phytomass* (decreasing steadily), and human appropriation of plant growth (a disturbingly large percentage of all plant growth).

1. *Agricultural land use.* We get a rough approximation of the extent of historical agricultural land use from Box 5-1, "World agricultural land use, 1700–2000." Total land used for agriculture has increased by perhaps sixfold just since the beginning of the Industrial Revolution and occupies more than one-third of the total ice-free land area of the world. It is this expanse of agricultural land that supports all the additional domesticated biomass, both humans and their livestock.

Most of this land use was and is for livestock agriculture. Obviously, the grazing land goes for livestock agriculture, and a good quantity of the cropland does as well; crops are grown and cultivated then fed to livestock animals. Cornell ecologist David Pimentel estimated that more than half of all the grain grown in the US is fed to livestock.[31] More than two-thirds of the cropland in the contiguous forty-eight states of the US is for livestock feed

---

31 *Cornell Chronicle,* "U.S. could feed 800 million people."

or exports or is idle.[32] Much of the other cropland is used for oilseeds (which partially support livestock agriculture), ethanol, biodiesel, cotton, or other nonfood crops, so less than 20 percent of cropland actually feeds people.

The inefficiencies of livestock agriculture have been well known to the public for decades.[33] The higher on the food chain one eats, the greater the energy lost: 80 percent of protein fed to poultry and 90 percent of protein fed to pigs is wasted in terms of the final food values for humans.[34] From the point of view of land use, it is vastly more efficient to eat plant foods directly than to feed plants to animals and eat the *animals.*

2. *Plant phytomass.* Another way of looking at the human impact on the plant world is by examining total plant biomass, called *phytomass.* With increasingly prolific agriculture spread all over the globe, it might seem that total phytomass would also have increased. But this isn't the case. Not only is the phytomass of the planet not keeping up with the increase in large animals, but during the past two thousand years it has *decreased* substantially.

According to one estimate from Vaclav Smil, total phytomass was more than one thousand gigatons of carbon (Gt C) five thousand years ago but by the year 2000 was only 550 Gt C, having decreased by as much as 45 percent in the past two thousand years and more than 16 percent just since 1900.[35] More recently, Karl-Heinz Erb and his colleagues reached an even more dramatic conclusion. Without human land use, total phytomass would be 916 Gt C, but today there is only about 450

....

32 Merrill and Leatherby, "Here's How America Uses Its Land."
33 First publicized by the best seller from F. M. Lappé, *Diet for a Small Planet.*
34 Smil, *Enriching the Earth,* p. 165, Figure 8.4.
35 Smil, "Harvesting the Biosphere" (2011), p. 616.

Gt C of phytomass on the planet, *less than half as much.*[36] These phytomass decreases have come largely because of deforestation and desertification, both of which are mostly due to overgrazing by cattle, though we can also blame logging and human habitat expansion.

3. *Plant growth and net primary production (NPP).* A third way of looking at human influence on the plant world is to investigate how much plant matter humans use today.

A number of people have tried to answer this question by looking at what biologists call *net primary production* (NPP). *Primary production* is mostly plant growth (plus some algae and bacteria), called "primary" because it is organic matter produced directly from inorganic compounds in soil and water.

How much of NPP is appropriated by humans each year? Research and analysis of this question has been underway for more than three decades. A pioneering 1986 study by Vitousek et al., gave three different estimates of about 4 percent, 31 percent, or 39 percent of total land-based NPP appropriated by humans[37] (see Box 5-2). Aquatic NPP appropriated was less, only about 2 percent of the total. A more recent review article found that estimates ranged from 3 to 55 percent of total NPP appropriated by humans, depending on assumptions and methodology, though most estimates ranged from 20 to 40 percent.[38]

These figures vary widely because all the researchers have a problem reaching a shared definition of exactly what it means for NPP to be "appropriated" by humans. The low figure from the Vitousek article only included the actual amount *consumed*

36 Erb et al., "Unexpectedly large impact of forest management and grazing."
37 Vitousek et al., "Human Appropriation of the Products of Photosynthesis."
38 Haberl, Erb, and Krausmann, "Global human appropriation of net primary production," Table 1.

by humans or their livestock, about 4 percent of land NPP and 2 percent of aquatic NPP, which doesn't sound that bad. But it's clear that humans have a broad effect on the plant world which goes far beyond direct consumption. If we chop down a forest, create a pasture, and graze cattle on it, then we have altered the whole ecology of the area, including the other animals that could survive there. The amount of NPP *co-opted* in such ways by humans for their own purposes is nearly one-third of the total NPP.

And what happens if the pasture is overgrazed and turns into a desert? In this case, the NPP of the area drops to about zero. Even if humans aren't currently consuming any NPP from these desert areas, humans have dramatically affected the area. And if we include the *potential* NPP lost (in a world in which humans never existed), the percentage of NPP co-opted by humans is even greater. The land turned into desert by various human activities (from acid rain to urbanization) throughout history is enormous: twenty million square kilometers, an area greater than the agricultural cropland now in production.[39] This NPP was destroyed by human intervention but cannot now be restored in any timely fashion. While all these figures vary considerably depending on their assumptions and what exactly they try to measure, they all agree on one thing: outsized human intervention in NPP has altered the basic botany of the planet.

----

39 Kovda, "Soil Reclamation and Food Production," pp. 160–161.

Description	Amount	Percentage
Consumed	5.2 Gt	3.9 percent of actual NPP
Co-opted	40.6 Gt	30.7 percent of actual NPP
Co-opted or degraded	58.1 Gt	38.8 percent of potential NPP

Box 5-2. Vitousek et al. estimates of human appropriation of land-based NPP

Source: Vitousek et al., "Human Appropriation of the Products of Photosynthesis." Calculated based on tables 1, 2, 3, and 4.

Total actual land based NPP = 132.1 Gt; Total potential NPP = 149.6 Gt

Percent of aquatic NPP consumed, co-opted, or degraded is less, estimated at 2.0 Gt co-opted / 92.4 Gt total NPP = 2.2 percent.

1 Gt (1 billion tons) = 1 petagram = 10^{15} grams

The Green Revolution

Of course human beings have used technology in agriculture to get the food they need; T for *technology* is the final variable in the IPAT equation. For the past fifty years, the Green Revolution has dramatically increased food production on the planet. May we assume that technology will solve everything? You can probably guess the answer to that question, but it's worth spelling out.

What we call the Green Revolution is the widespread use of agricultural techniques pioneered by Norman Borlaug. The Green Revolution likely prevented widespread famines in the 1970s, when population trends seemed to point toward disaster, and Borlaug later justly received the Nobel Prize for his work. The Green Revolution dramatically improved crop yields by focusing on new, high-yield grain varieties, in combination with irrigation, pesticides, and fossil fuel energy, to provide greater quantities of food on the same amount of land.

One unexpected consequence of the Green Revolution was that as more and more grain was produced, it became *cheaper*, and this stimulated demand for still more grain. Instead of dealing with the population problem, we simply decided we had "solved" the problem and continued much as we had before. Instead of using the Green Revolution's efficiency to conserve resources or feed the hungry, the economy funneled the grain surpluses to the livestock industry to produce cheap meat. Human population continued to increase, and livestock population grew even faster.

Our expectation is that as we get more efficient technology, our resource requirements will decline. But technological efficiency often *increases* resource consumption, a phenomenon known as the "rebound effect" of efficiency gains or the Jevons paradox, named after a nineteenth-century English political economist named Stanley Jevons. In *The Coal Question*, Jevons pointed out an apparent paradox. Greater efficiency in the use of coal (via the Watt steam engine) also made coal use cheaper, but this stimulated greater demand for coal. The resulting increased demand for coal vastly outweighed the efficiency gains of the steam engine, and total coal usage *increased*.

The same thing applies to the Green Revolution. By making food production more efficient, the Green Revolution has also made food cheaper, facilitating both an increase in human population and an increase in livestock population to absorb the surplus grains. Much of the Green Revolution's output is being funneled into livestock. Meat consumption has dramatically increased both in absolute and per-capita terms since the Second World War—a highly profitable but tremendously inefficient way to provide food for humans. Most of the plant nutrition fed to animals is lost in the process of being turned into meat.[40]

40 Weis, *The Ecological Hoofprint*, p. 115, is one of numerous authors pointing this out.

The Green Revolution had another unintended result. While it does economize on land requirements, it makes great use of irrigation water and energy resources—which are themselves limited and depleting. In the 1970s, both water and energy were relatively cheap, but this is much less true today. The Green Revolution doesn't so much *decrease* the resource requirements of growing food as it *shifts* them, from total land to total water and energy. Technology can postpone or shift limits but cannot permanently evade them.

Norman Borlaug, the Nobel Prize winner who was responsible for much of the Green Revolution's technology, was considerably more sanguine about the Green Revolution than many of its promoters. He warned that this technology was only a temporary reprieve so that we could have time to deal with what he called the "Population Monster."[41] Now, human population has approximately *quadrupled* since Borlaug's birth and *doubled* since he received the Nobel Prize in 1970. Livestock biomass has increased even faster, with per-capita consumption of meat *doubling* in the developing countries just between 1980 and 2002.[42] Instead of addressing the population problem, food surpluses are being used to support greater numbers of people feeding on even greater quantities of livestock animals.

We have ignored Borlaug's warning, and now we must squarely face the consequences. We have delayed but not prevented the "population bomb" famously (though prematurely) predicted by Paul Ehrlich.[43]

....

41 Borlaug, "Nobel Lecture," December 11, 1970.
42 Thornton, "Livestock production."
43 Ehrlich, *The Population Bomb*.

Conclusions

We started out with the IPAT equation but found that this equation leaves out quite a few relevant details. Most important, how do livestock figure into this equation? Should they just be subsumed under the "affluence" category, or what? What about the impacts on biodiversity, which is crashing under the strain of human and livestock numbers? Why are we even concerned about human population if livestock population (by biomass) is twice as large? And what about the consequences of technology and the Green Revolution, which instead of decreasing resource use has increased it, both because of the "rebound effect" and because of increased water and energy usage?

We don't have enough information—yet—even to approximate the sustainable human population. This figure depends on all the other resources and assumptions taken together. But we already know enough to make this fundamental point: *humans overwhelmingly dominate the biology of the planet.* This dominance has dramatically increased in the twentieth century alone. This has numerous environmental implications for wildlife, plants, and even the climate, which we will explore in coming chapters.

While megafauna biomass has exploded since 1500, this is mostly because of the increase in domesticated megafauna (our livestock), while wild megafauna are in catastrophic decline. (Smaller creatures are also affected, as we will see below.) In the meantime, plant phytomass has substantially declined, mostly due to deforestation and desertification. Agricultural productivity has compensated for some loss of phytomass, but the overwhelming consequence of human domination of the world has been a sharp decline in total plant matter. So far, the consequences of this destruction have fallen mostly on the wild plants and animals of the planet, but a day of reckoning for humans is fast approaching—if it is not already here.

CHAPTER SIX

ARE WE THE NEXT DINOSAURS?
Biodiversity and mass extinctions

As we've seen, while the total biomass of land megafauna has soared to new heights in recent centuries, this rise is due to dramatic increases in numbers of humans and their livestock. As humans and their livestock overrun the planet, we are crowding out wild species—and driving many toward extinction.

Human-caused extinctions have played an important role in human history. In prehistoric times, extinctions shaped the broad contours not only of our biophysical environment (the plants and animals around us), but also of our societies, in ways almost invisible to most of us.

Today's rate of extinctions (which is one hundred out of every million species, per year) is one thousand times greater than the normal "background" rate of extinction (which is only 0.1 species/million species/year).[44] Some might argue that these numbers, on their face, don't pose an immediate problem. At this rate, it would take a century to drive ten thousand species out of a million to extinction—about 1 percent of all species.

---

44 De Vos et al., "Estimating the normal background rate of species extinction."

We wouldn't reach a 50-percent extinction rate—something on the scale of a "mass extinction event"—for five thousand years.

However, we can't quantify the ecological contribution of each species just by counting its numbers. There's quite a bit we don't know about mass extinctions, such as what "pulls the trigger" on mass extinctions, how long they take, and which species have the best chance of surviving.

A species can be condemned to extinction even though some of its members are still alive, due to habitat degradation or other problems.[45] Mammoths had almost completely disappeared by about ten thousand years ago due mostly to human influences, but isolated pockets of mammoths survived in Siberia as recently as five thousand years ago.[46] The trigger for the end-Cretaceous extinction—the one that killed the dinosaurs—happened in a single day, when an asteroid collided with Earth, though the extinction process may not have been completed for several thousand years afterward. The late Devonian extinctions (beginning about 372 million years ago), by contrast, lasted ten to fifteen million years.

The issue of extinction raises ethical, aesthetic, and practical issues. Do we have the right to render elephants, or any other species, extinct? Would we really want to live in a world without these wild species? From a purely practical perspective, if we wittingly or unwittingly bring about a sixth great extinction event, could we ensure that humans will not be one of the victims?

Past extinctions and "prehistoric overkill"

The biosphere has been remarkably stable for millions of years. These periods of stability have been interrupted by five

45   Barnes et al. "Dead clades walking."
46   University of Adelaide. "Humans hastened the extinction of the woolly mammoth."

great extinction events since the "Cambrian explosion" of multicellular life on Earth about 541 million years ago. These major extinctions led to a catastrophic decrease not only in the total amount of life on the planet, but also in the total number of species. The end-Cretaceous extinction (about 66 million years ago), initiated by an asteroid hitting the earth, killed off the non-avian dinosaurs and about three-quarters of all species. The worst major extinction, the end-Permian extinction about 251 million years ago, was nearly fatal for life on Earth: most living things died and an estimated 95 percent of all species went extinct.

Might we be on our way to a sixth mass extinction today? If so, future paleontologists would probably date its onset to the most recent killing of large mammals in prehistoric times, beginning about fifty to sixty thousand years ago.

The widespread extinction of large animals that occurred around this time was almost certainly primarily due to overhunting by humans. Modern humans, *homo sapiens*, originated in Africa only about two hundred thousand years ago. Sometime about sixty thousand years ago, humans migrated out of Africa to Asia, Europe, Australia, and then North and South America. By the time they left Africa, humans were skilled hunters, but large mammals on other continents were biologically unprepared.

As a result of human hunting skills and human migration, species in areas of human migration started going extinct following their arrival. Human migration to the Americas helped bring about extinction to mastodons, mammoths, saber-toothed tigers, glyptodons, giant sloths, and American species of both the horse and camel. This same extinction pattern seems to have followed humans whenever they migrated to new areas of the world. Large animal extinctions in Asia, Australia, North America, South America, and Madagascar and other islands directly followed the migration of homo sapiens to those areas (see Box 6-1, "Prehistoric large mammal extinctions and human migration"), a process called "prehistoric overkill."

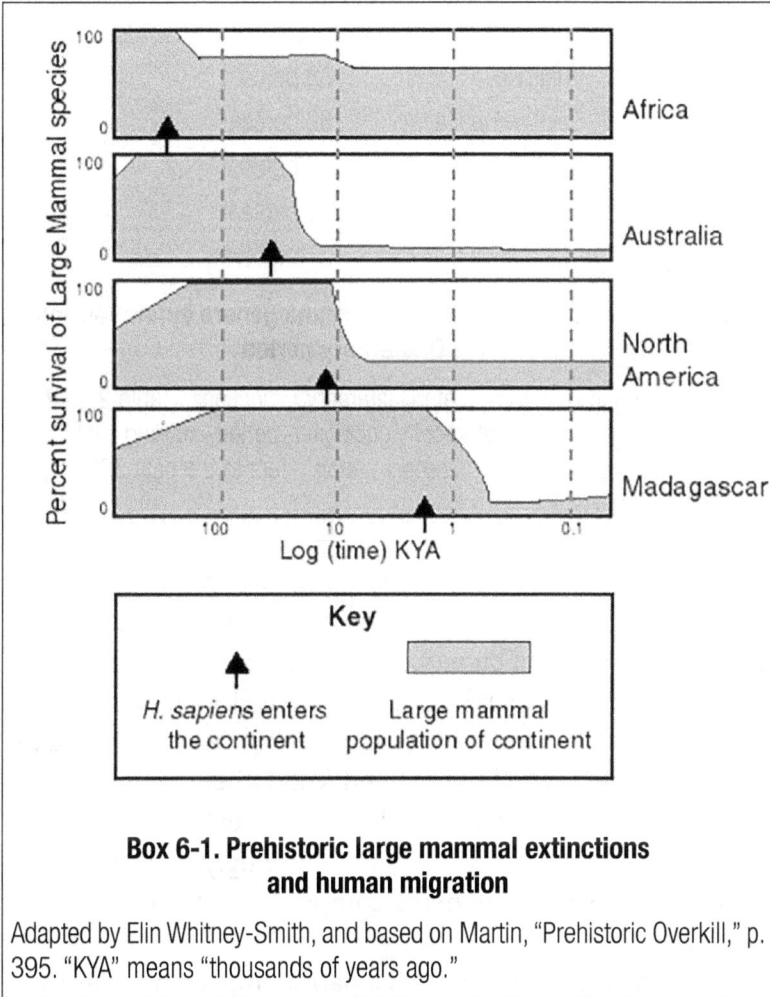

Box 6-1. Prehistoric large mammal extinctions and human migration

Adapted by Elin Whitney-Smith, and based on Martin, "Prehistoric Overkill," p. 395. "KYA" means "thousands of years ago."

Ironically, the one key exception to this pattern is in Africa, where homo sapiens began. Large mammals in Africa evolved at the same time that humans were developing their still-primitive hunting skills over many thousands of years, and they likely learned to be wary of humans. Such notable African species as the elephant, rhinoceros, hippopotamus, giraffe, and lion survived into the twenty-first century. Species extinctions were much greater on other continents (see Box 6-2, "Large animal extinctions due to prehistoric overkill").

Australia	88 percent
South America	83 percent
North America	72 percent
Eurasia	35 percent
Africa	21 percent

Box 6-2. Percent mammalian megafauna genera extinct during the late Quaternary period

Source: Koch and Barnosky, "Late Quaternary Extinctions," table 2, p. 219. The Quaternary is the most recent geological period, starting about 2.6 million years ago. The late Quaternary period refers to the past 500,000 to 1,000,000 years or so.

Paul Martin developed the key arguments for the "prehistoric overkill" hypothesis.[47] Again, the extinctions were sudden and followed the spread of humans. Large mammals were decimated, reflecting human hunters' selection, but smaller species survived. Large mammals mostly survived in Africa, where megafauna had co-evolved with humans and knew them as dangerous. Some species were only indirectly rendered extinct; saber-tooth tigers may have gone extinct not because they were hunted by humans, but because humans competed with them for food supply.

Other factors may have had a partial impact in some areas. Climate change, in the form of a global cooling event about eleven thousand years ago (the "Younger Dryas"), may have been a factor in the American extinctions. (On the other hand, the "Younger Dryas" itself may have been indirectly *caused* by humans due to human overhunting of ruminant species

47 Martin, "Prehistoric Overkill," pp. 354–403.

that produced low-level methane emissions and warmed the planet.[48]) But in other regions, such as in Australia and the Pacific Islands, extinctions occurred without climate change. And even in America, species such as mammoths and giant sloths had survived many previous ice ages over the previous million years or so. The prehistoric overkill hypothesis is today increasingly accepted as the primary cause of these extinctions beginning about fifty thousand years ago.

The origins of hunting, warfare, and agriculture

If "prehistoric overkill" caused these extinctions, why and how did human hunting of animals develop in the first place? Humans aren't well adapted to eat meat, and meat-eating causes numerous health problems.[49] Our closest primate ancestors and relatives (*australopithecus* and modern relatives such as apes, bonobos, and chimpanzees) are largely or entirely vegetarian. And yet today, we have huge industrial livestock systems and *most* of the large animals on the planet wind up killed as food for humans. How and why did humans come to hunt animals and eat meat?

Steven A. LeBlanc, an American archeologist formerly at Harvard, argues that hunting and warfare co-evolved out of the same technology and social organization.[50] Humans evolved in a world with many large animals. Their behavior started out as self-defense against predators, which were a key limit on human population. But as humans became more successful against predators, they switched from defense to offense, utilizing animals for food. With increasing tool use, humans increasingly relied on hunting for food.[51]

An important side effect of hunting was that it established

48 Smith et al., "Methane emissions from extinct megafauna."
49 See chapter 17.
50 LeBlanc and Register, *Constant Battles*, pp. 90–91, 97–98.
51 LeBlanc and Register, *Constant Battles*, p. 98.

war as a successful strategy. While predators ceased to be as dangerous, other humans became a danger, and intergroup conflict arose. The same skills that made humans so dangerous to large animals—such as communication, coordination, and weapons—*also* made them more dangerous to other humans and effective in warfare. For most of human history, the tools of the hunt were also the tools of warfare.[52] Constant battles and violent death then became a leading cause of death in primitive peoples—much more common, on a per-capita basis, than it is today. The number of violent deaths (per capita) in modern times, despite highly destructive events such as world wars and the Holocaust, is *vastly* less than it was in more primitive hunter-gatherer societies. This was most famously pointed out by Steven Pinker in his bestselling book *The Better Angels of Our Nature*, but other scholars agree on this general point, such as Lawrence Keeley, Steven LeBlanc, Katherine E. Register, Iain Provan, and Peter Turchin.

It was economically advantageous for a tribe to hunt animals for food rather than gather plants or cultivate crops. Hunting had a higher *energy return on energy invested* (EROEI—see chapter 14). The energy gained through eating meat was much greater than the effort expended to acquire it[53] and more advantageous than using the same effort to gather plants or cultivate crops, considering the whole process of acquiring, preparing, and eating food and factoring in risks and other benefits (such as defensive protection against wild animals).

If hunting returned *less* energy than gathering plant foods or scavenging (per unit effort), it seems unlikely that humans would have taken it up. They likely would have stuck with the tried-and-true methods that gave them greater caloric return.

....

52 LeBlanc and Register, *Constant Battles*, p. 91.
53 Steinhart and Steinhart, "Energy Use in the U. S. Food System,"
    p. 312, figure 5, give the ratio of energy gained to energy
    expended for hunting and gathering as about 10:1.

It is implausible that this switch occurred because of some nutritional advantage of animal foods. Our hominid ancestors diverged from gorillas and chimpanzees about six million years ago, but modern humans are still remarkably similar to gorillas.[54] Modern gorillas eat a mostly vegan diet, based on leaves, stems, and herbs. If gorillas are fed a more concentrated, calorie-dense diet, they start to develop heart disease and obesity—as do modern humans, as we will see in chapter 17. Zoo gorillas fed a concentrated diet were switched to a more natural vegan diet, full of their wild bulky food, and their health rapidly improved.[55]

On the other hand, this means that gorillas eat twenty-five to forty pounds of food per day. Gorillas in the wild travel long distances and spend quite a bit of time gathering this food.[56] Our hominid ancestors likely experienced similar demands. The caloric concentration of foods due to eating meat from hunted animals may have presented health problems, but these would have only affected the individual. They wouldn't have affected the tribe, as they mostly would have started to affect individuals when they were well past reproductive age.

The benefits of meat-eating were *economic* advantages. The tribe that hunted for its food spent less time both gathering and consuming food. They had more time to devote to other tasks, such as developing more advanced spears, discovering new uses for fire, or developing social networks. Tribes relying on meat as food were able to outcompete tribes that (for whatever reason) did not; they had more spare time, and they were able to compete more effectively in warfare.

As hunting was perfected, it provided an ample supply of previously untapped food energy. Before humans came along, being a large animal conferred an obvious evolutionary advantage. Larger animals were harder to kill. Mammoths,

---

54 Mosher, "Gorillas More Related to People Than Thought."
55 Case Western Reserve University. "Gorillas go green."
56 "Gorilla Feeding," by Gorillas-World, February 6, 2014.

sloths, and American camels were large. But humans figured out—with tools, communication, and cooperation—how to kill these large animals. Hunting skill against animals larger than humans was a major evolutionary breakthrough and gave humans a tremendous store of potential food unique in the planet's evolution. With the advent of humans, being a large animal was suddenly an evolutionary *disadvantage*, as humans could easily find you. It is easier to find elephants than squirrels.

If hunting was such a great strategy, why was it mostly abandoned about ten thousand years ago? Some modern hunter-gatherers spend much less time "working" than modern Americans. Why would prehistoric human beings have voluntarily gone from a system in which they might have needed to work four hours a day or fewer, on average, to an agricultural system requiring long hours and intensive labor?

The best explanation is that necessity was the mother of invention: the hunter-gatherer existence wasn't sufficient to support the growing human population.[57] Due to "prehistoric overkill," humans had run out of easy hunting targets, and with rising human numbers, hunting and gathering became increasingly difficult—which probably coincided with the time frame when humans began to widely adopt agriculture. The very success of hunting led to its demise as a successful strategy.

At that time, the human population was about three million, less than 1/2000 of our current population. Even with this small population, humans had driven many large animal species extinct. As a practical matter—unless we are talking about eliminating more than 99.9 percent of the human population—there is no chance that humans can ever return to a hunter-gatherer means of existence.

In retrospect, agriculture was perhaps the most successful cultural adaptation of all time. In a relatively short time, human

57 Cohen, *The Food Crisis in Prehistory*. This thesis was rather daring in 1977, but now seems to be increasingly accepted.

population exploded to more than one hundred times its previous level. Hunting animals thereafter was mostly a sport, practiced only by elites able to chase all the other humans off the remaining land harboring the few remaining species.

Modern Extinctions

Beginning about the year 1500, another round of human-caused extinctions began, initially due to overhunting but later to habitat destruction and other human activities. Writers such as E. O. Wilson, Elizabeth Kolbert, and Anthony Barnosky have all written eloquently about modern extinctions that spiked in just the past five hundred years, increasing further after the Industrial Revolution and accelerating in the twentieth century. The passenger pigeon, the Tasmanian tiger, the quagga, the dodo, several species of moas, the Javan tiger, the Carolina parakeet, and the ivory-billed woodpecker have all become extinct.

Smaller creatures than the megafauna are also affected, from insects to intermediate-size animals not quite large enough to fall under the "megafauna" category. In fact, perhaps even more alarming than the decline in large mammals is the threat to invertebrate insects and the species that rely on them for food. Many people (including this author) have noted anecdotally a decline in "windshield bugs," insects splattered on a car windshield during long car trips. Scientists have published a number of alarming reports documenting the recent decline of bee diversity,[58] as well as 80-percent declines in insect populations in Germany.[59] One study of Puerto Rico's Luquillo rainforest found that arthropods (invertebrates, including many insects, that have external skeletons) had declined from ten to sixty times in Puerto Rico compared to their 1970s densities, with parallel declines in lizards, frogs, and birds that eat

....

58 E. g., Razo-León et al., "Changes in bee community structure."
59 Vogel, "Where have all the insects gone?"

arthropods.[60]

Noted insect ecologists Francisco Sánchez-Bayo and Kris Wyckhuys glumly concluded that "over 40 percent of insect species are threatened with extinction."[61] These declines will critically affect both populations of insect-eating vertebrates, such as birds, and any plants dependent on insects to pollinate them. The authors blame habitat destruction, pollution, agricultural intensification, and pesticides; they acidly comment: "Unless we change our ways of producing food, insects as a whole will go down the path of extinction in a few decades.... The repercussions this will have for the planet's ecosystems are catastrophic to say the least."[62]

Conservation scientists summarize the causes of these extinctions, all related to human activity, from greatest to least impact, with the acronym HIPPO: Habitat destruction, Invasive species, Pollution, Population growth, and Overhunting (including fishing).[63] As a practical matter, it's difficult to see a way that we can deal with modern extinctions without dealing with livestock agriculture. Agriculture is the leading human use of land, thus the cause of most habitat destruction. Human overpopulation is significant also, but the effects of human overpopulation alone are dwarfed by the overpopulation of *livestock*, simply because livestock occupy so much more land.

Agriculture also is a leading source of pollution; it has led to plastic pollution from deep-sea fishing (discarded

60 Lister and Garcia, "Climate-driven declines in arthropod abundance." "Ten to sixty times" means that the insect populations were ten to sixty times greater in the 1970s than they were when this study was done, so the declines were to about 1.6 percent to 10 percent of previous abundance.

61 Sánchez-Bayo and Wyckhuys, "Worldwide decline of the entomofauna," p. 8.

62 Sánchez-Bayo and Wyckhuys, "Worldwide decline of the entomofauna," p. 27.

63 Wilson, *Half-Earth*, pp. 57–58.

fishing nets),[64] nitrogen pollution from nitrogen fertilizers, and pesticide contamination. Researchers have warned that we have already exceeded the planetary boundary for "novel entities"—manmade pollution such as plastics and fertilizers— in the environment.[65] Agriculture, as commonly practiced, has led to significant decline in overall biodiversity. By design, it clears natural diverse landscapes and substitutes a narrow range of plant species—often, just one species—favored as human (or livestock) food. The fewer species and less biodiversity an ecosystem has, the more susceptible it becomes to chance weather events, new diseases, or other random events. Some agricultural monocultures (vast amounts of land dedicated to a single crop) have already suffered disastrous consequences. The potato monoculture in nineteenth-century Ireland resulted in a disastrous famine when a new disease struck the potato, with a resultant drop in Ireland's human population of about 15 to 20 percent.

Nearly ubiquitous pollution, declines in biodiversity, pesticide use, and habitat destruction are key factors not only in degrading wildlife environment, but also in degrading *our own* environment. All this not only promotes wildlife extinction, but also inadvertently promotes our *own* extinction.

Biosphere 2 as a cautionary tale

From a purely *practical* point of view, what difference does extinction make? Would humans be adversely affected if elephants—or for that matter most or all wild species of plants, animals, insects, and fungi—were to become extinct?

The Biosphere 2 experiment from the 1990s demonstrates that we have real reasons for concern.[66] Biosphere 2 was launched

64 Laville, "Dumped fishing gear."
65 Persson et al., "Outside the Safe Operating Space."
66 A sympathetic account from one of the participants is Poynter, *The Human Experiment.*

by billionaire philanthropist Edward Bass and systems ecologist John Allen in the Desert Southwest near Tucson, Arizona. The idea was to study humans and an array of plant and animal species in a controlled environment. One of the applications was to see if an artificial but balanced environment could be designed for use on other planets like Mars.

A team of eight humans were sealed up in an artificial closed environment, along with 3,800 different species of vertebrate and invertebrate species and various plant species in various habitats over about three acres. Species also included domestic animals: goats, pigs, chickens, and fish. It was designed to house everyone over a space of two years. The habitats included a small "ocean," a mangrove wetland, a grassland savanna, and an agricultural area.

The basic idea was ingenious, but the results were disturbing. In Biosphere 2, where the eight humans were obviously the dominant species, the environment became unlivable. Maintaining an atmospheric balance was difficult. Carbon dioxide levels soared at one point to 7,400 parts per million (ppm), oxygen levels fell, and eventually oxygen had to be pumped in from the outside. Cockroach populations exploded as well as populations of an inadvertently added ant, the "crazy ant" (*Paratrechina longicornis*), which quickly become the dominant ant.[67] The animal diversity in Biosphere 2 was "severely reduced," and all the known pollinators, such as bees and hummingbirds, disappeared.[68] The humans could grow food, and their food yields were quite high, but they couldn't grow enough to satisfy their hunger.

Some of these problems could likely be corrected were this experiment repeated; a construction defect turned out to be the root cause of falling oxygen levels. Despite the project's collapse,

....

67 Wetterer, Miller, Wheeler, and Olson, "Ecological Dominance by Paratrechina Longicornis."
68 Wetterer, Dunning, Yospin, and Himler, "Invertebrate Diversity and Ecology in Biosphere 2."

the Biosphere 2 experiment did generate some interesting scientific results. But Biosphere 2 highlights the fact that *a seemingly sustainable combination of species and environments can become unlivable.* What will happen in the real world now when we are degrading our habitat and altering the species makeup of the entire planet on an immense scale?

In addition to degrading the habitat of wild creatures both large and small, we are also degrading our *own* habitat, that interconnection of life that supports our existence in countless ways. *Our* habitat might be degraded to the point where it cannot support us any longer. Paleontologist Henry Gee pessimistically argues that for these and similar reasons, humans are *already* condemned to go extinct.[69] If we are to avoid this outcome, we need to act decisively and soon against these trends.

One of the adverse phenomena in Biosphere 2, the disappearance of insect pollinators, is already manifest in *our* biosphere. And we haven't even mentioned the disappearance of non-insect pollinators like bats,[70] or of such seemingly mundane species as snails.[71] Nor have we discussed amphibian extinctions; while declines of mammals and bird species "paint a gloomy picture, it is considerably worse for amphibians."[72] What we face is not just a decline in a few wild species, but disastrous, across-the-board declines almost everywhere.

At the present time, resuscitation of extinct species is almost impossible.[73] We are conducting a gigantic, one-time, irreversible experiment on Earth's biosphere. The burden of proof should rest on those who advocate or cause actions that destroy habitat and otherwise contribute to the causes of species extinction to

....

69 Gee, *A (Very) Short History of Life on Earth*, p. 188.
70 Platt, "Are Bats Facing a Hidden Extinction Crisis?"
71 Platt, "Snails Are Going Extinct."
72 Bishop, P. J., "The Amphibian Extinction Crisis."
73 See discussion in Barnosky, *Dodging Extinction*, chapter 7, "Resuscitation."

prove those actions are safe, not on environmentalists to prove they are unsafe.

Conclusions

In prehistoric times, human overhunting led to extinction of most of the wild megafauna in North and South America. In modern times, extinction is being driven by numerous human activities, most notably agricultural land use that deprives wild animals of habitat. The root cause of these modern extinctions is the continual expansion of the human sphere at the expense of everything else.

Sustainability is more than just creating an inventory of all the natural resources on the planet and making sure there is enough for all the world's *humans*. The Biosphere 2 experiments seem to have highlighted quite well the dangers of such an approach for humans. We are playing with fire when we reduce entire environments to dust and drive scores of species to extinction.

CHAPTER SEVEN

HIDDEN DIMENSIONS OF CLIMATE CHANGE

Forests, cattle, and the atmosphere

Earth's climate is warming, human activities are the primary cause, and there is overwhelming scientific consensus about these realities.[74] There is also general agreement that fossil fuels are heavily implicated, and that therefore, fossil fuel burning needs to be restricted somehow.

Do we need to say more? What can we contribute to this discussion?

Current state of public discussion on climate change

Unlike some of the problems addressed in this book, the *reality* of climate change is a problem not only well established in scientific literature, but also for which there is significant public awareness. Popular books such as those by renowned climate

---

74 *Global Climate Change* (website), "Scientific Consensus: Earth's Climate Is Warming."

scientist James Hansen[75] and former Vice President Al Gore[76] provide an excellent window into the scientific explanations. We are well beyond the stage at which we should still be creating awareness of the problem; informed public discussion has shifted to what we can do to address it.

That public discussion has concentrated on *solutions*: dealing with the politics of the situation[77] or implementing various technical approaches.[78] In this chapter, we want to revisit the problem of the *causes* of climate change.

Scientific discussion, while not questioning the heavy involvement of fossil fuels, raises another alarming possibility. Land use *also* contributes to climate change more significantly than generally believed. Popular discussion hasn't caught up with science here, though Jonathan Safran Foer[79] provides the best popular overview of the case for the significance of land use.

For practical purposes, "land use," in this case, is a way of saying "livestock agriculture"—not because there aren't other uses of land that also affect climate, but because livestock agriculture is overwhelmingly the dominant human use of land. Livestock agriculture, when you figure that it not only uses vast quantities of land, but also burns a substantial amount of fossil fuels, may actually account for about half of our climate problem. That discussion of climate has inadequately factored in land use means we have significantly *underestimated* both the human contribution to climate change and the role of livestock.

Measuring land-use impact on climate change

What is the relationship between climate change and our gigantic disruption of the biosphere through land use and deforestation?

....

75 Hansen, *Storms of My Grandchildren.*
76 Gore, *Our Choice.*
77 E. g., Klein, *This Changes Everything.*
78 E. g., Hawken, ed., *Drawdown.*
79 Foer, *We Are the Weather.*

Most climate change assessments focus on *direct* emissions of greenhouse gases, such as coal-fired power plants putting CO_2 into the atmosphere. They disregard or discount the *indirect* contributions of CO_2 to the atmosphere resulting from human disruption of the natural carbon cycle.

The natural carbon cycle constantly cycles greenhouse gases into and out of the atmosphere, as well as into and out of the ocean and soil. Normally, this carbon cycle should be roughly in balance; greenhouse gases are pulled out of the atmosphere (e.g., through photosynthesis) at approximately the same rate that they enter the atmosphere (e.g., through respiration). But suppose that, through human actions, plant photosynthesis is inhibited, or animal respiration is increased, or both? Obviously, this could disrupt the carbon cycle and lead to a buildup of CO_2 in the atmosphere. This wouldn't be a *direct* human contribution of CO_2 such as burning coal, but an indirect contribution via altering natural processes.

The quantities involved in the natural carbon cycle are *immense*. Carbon is cycling in and out of the soils, the oceans, and the atmosphere annually at a rate many times greater than we are contributing carbon to the atmosphere by fossil fuels. There is more than twice as much carbon in the soils, and more than fifty times as much carbon in the oceans, as there is in the atmosphere. A relatively small disturbance in this natural cycle could have tremendous repercussions in terms of the climate.

We can measure or at least estimate direct greenhouse gas emissions through the burning of coal, but it is more difficult to measure the impact of disturbances to the natural carbon cycle. Our disruptions of the biosphere not only affect the climate right now, but they also already *did* even before the Industrial Age and before humans started using fossil fuels on a large scale.

The role of pre-industrial agriculture

Livestock agriculture has been practiced for about ten thousand years. It started well before the Industrial Revolution. If livestock agriculture and human land use were responsible for climate change, shouldn't we *already* have observed evidence of climate change prior to the Industrial Revolution?

Actually, there *is* such an effect on climate from pre-industrial agriculture. William F. Ruddiman, a paleoclimatologist and Professor Emeritus at the University of Virginia, wrote a pioneering book *Plows, Plagues, and Petroleum*. Ruddiman shows that humans *already* had an impact on the climate not only prior to the Industrial Revolution, but also prior to recorded history.

Today's CO_2 levels are well over 400 ppm (parts per million). Most people, including most climate scientists, assume that natural levels of CO_2 today would be what the levels were in 1700 before the Industrial Revolution, namely 280 ppm. Ruddiman argues that the *true* natural level, the level that CO_2 would be at without human activities, is 240 ppm. In other words, even before the Industrial Revolution, humans had already raised atmospheric CO_2 by about 40 ppm.

Ruddiman derives the figure 240 ppm by examining the long-term "ice age" cycles. Because of slight variations in the earth's orbit and tilt and the dominance of most of the earth's land mass in the northern hemisphere, the planet's climate has alternated between ice ages and warmer periods over the past few million years. (These variations were described by Serbian astronomer Milutin Milankovitch early in the twentieth century.) As the ice ages have come and gone, CO_2 and CH_4 levels in the atmosphere have oscillated up or down. Rising or falling CO_2 levels are both an effect of the warming or cooling of the oceans and a cause of further warming (or cooling), which in turn also affects CH_4 levels. Due to these oscillations, we should see a new ice age every 100,000 years or so.

Beginning about ten thousand years ago and continuing today, the planet should be in a long-term *cooling* trend. CO_2 levels were gradually falling and should have reached 240 ppm by the year 1700, but in fact they reached 280 ppm. CH_4 levels were also declining toward 450 ppb (parts per billion) but reversed course about five thousand years ago, and by 1700 they were back up to over 700 ppb.[80]

Something was happening even before the Industrial Revolution to add CO_2 and CH_4 to the air. Most likely, it was human agricultural activity: namely, clearing forest land and burning the trees, both of which added CO_2 to the air. Rice cultivation and livestock use also increased, both activities increasing CH_4. We can't account for this by "natural variation" because no comparable increases, for *either* CO_2 or CH_4 (determined through analysis of ancient ice cores), occurred during any of the six previous interglacial periods. Ruddiman concludes: "Based on these failures, a strong case can be made that natural explanations (of any kind) have been falsified."[81] Ruddiman's thesis has gained general acceptance in the scientific community, and in 2010 Ruddiman was awarded the Lyell Medal of the Geological Society of London for his work.

Ruddiman adds an interesting twist here that further strengthens his case. While pre-industrial CO_2 levels generally increased leading up to 1700, there were "pauses" in increases from about 200 to 600 CE and after 1300. These pauses correspond to widespread outbreaks of deadly disease. Widespread plagues occurred in the ancient Mediterranean (200–600 CE), as well as from 1300 to 1500 in Europe (the "Black Death"), and from 1500 to 1800 in North America (plagues inflicted on the indigenous population by newly arrived Europeans). These plagues dramatically reduced human

---

80 Ruddiman, *Plows, Plagues, and Petroleum*, pp. 87, 77.
81 Ruddiman, *Plows, Plagues, and Petroleum*, p. 199; parentheses in original.

population, killing from 30 to 90 percent of the population in some places. The death of all these humans curtailed humans' clearing of the land, and the land began to revert to forest, thus absorbing some of the CO_2 that had been emitted and causing a fall in atmospheric CO_2. When the Europeans arrived in the New World, they brought plagues that killed 90 percent of the indigenous population. The resulting revegetation of 56 million hectares of uncultivated land caused a perceptible drop in CO_2 levels of about 5 ppm and a global cooling.[82]

Since 1700, we have seen an enormous upsurge in fossil fuel burning that has poured quantities of CO_2 into the atmosphere. But livestock agriculture has also dramatically increased, with grazing land about six times greater than it was in 1700. This leads us to the next question, which Ruddiman doesn't address: What has been the effect of the vastly increased modern agricultural system on the climate *since* the beginning of the Industrial Age?

"Livestock and Climate Change"

In 2006, the United Nations Food and Agriculture Organization (FAO) issued a report (*Livestock's Long Shadow*), that put greenhouse gas emissions from livestock agriculture at 18 percent of all human-caused greenhouse gas emissions. Since then, under pressure from the livestock industry, the FAO has revised this figure down to 14.5 percent.[83]But in November 2009, Robert Goodland and Jeff Anhang published an article for *WorldWatch* magazine titled "Livestock and Climate Change" (LCC). They argued that actual totals of greenhouse gas emissions from livestock agriculture were more than *four times greater* than the 2006 FAO estimate, at 51 percent of the

---

82 A. Koch et al., "Earth system impacts of the European arrival."
83 Goodland, "FAO Yields to Meat Industry Pressure on Climate Change."

total (Box 7-1, "'Livestock and Climate Change' estimates").[84] There has only been one attempt to respond to these specifics in a peer-reviewed setting, which was met with a vigorous response from the original authors.[85]

These estimates are *wildly* different. There is a major difference between 18 percent (or 14.5 percent) and 51 percent, as well as between the FAO's estimate in *Livestock's Long Shadow* of 7,516 Mt CO_2e (millions of tons of CO_2-equivalents) from livestock agriculture, versus LCC's estimate of 32,564 Mt CO_2e. So what's going on here?

There were three elements to the LCC article that set it off from the earlier FAO estimate: (1) better accounting, (2) methane, and (3) land use.

1. *Better accounting* (items 4 and 5 in Box 7-1). This element explains about one-third of the discrepancy. LCC said that the FAO had undercounted the total number of cattle, using data that was out of date or just wrong. LCC also looked at *misallocated* items: items such as waste, packaging, and medical care that were included in the FAO's estimate but should have been credited to livestock.

2. *The importance of methane* (item 3 in Box 7-1). It is well known that a disproportionate amount of atmospheric methane comes from cattle belching (which they emit while digesting the grasses they eat). But how warming is this methane? LCC argued that methane has a much greater warming effect than we previously thought; therefore, calculations for the amount of global warming resulting from methane emitted from cattle need to be increased.

....

84 The Goodland and Anhang estimate of greenhouse gases from livestock was 32.564 Gt CO_2e; the 2006 FAO estimate was 7.516 Gt CO_2e.

85 (a) Herrero et al., "Livestock and greenhouse gas emissions."
(b) Goodland, R. and Anhang, J. "Livestock and greenhouse gas emissions ... by Herrero et al."

The warming effects of a given amount of methane are expressed as CO_2-*equivalents*, that is, the amount of CO_2 needed to have the same warming impact. The warming effects of a given amount of methane are *much* greater than the warming effects of the same amount of carbon dioxide—on the order of one hundred times greater. But to add to the complications, methane breaks down relatively quickly in the atmosphere; it has a much greater warming impact but doesn't stick around in the atmosphere as long. The *half-life* of methane is about eight years, so that after about a decade, half of a given amount of methane will have broken down into something else (mostly, carbon dioxide).

The standard way of expressing the warming impact of methane is therefore to look at the warming impact over a specific time period. But what time period? The half-life of carbon dioxide is often considered to be about a century. Using a one-hundred-year time frame, the global warming potential (GWP) of methane is thirty-four times that of the same quantity of carbon dioxide if we include dangerous *climate carbon feedbacks*. But over a twenty-year time frame, the GWP is proportionally much higher, about *eighty-six times* that of equal amounts of CO_2, including the same climate carbon feedbacks.[86]

Climate carbon feedbacks are dangerous because warming can generate further warming. For example, the melting of snow or sea ice exposes the ocean surface below; the ocean absorbs more solar radiation than snow or ice would have, causing an increase in warming. Warming also leads to greater water evaporation, but water vapor itself is a powerful greenhouse gas that traps additional heat. If feedbacks such as these become too pronounced, the climate may pass a *tipping point* at which warming effects become self-reinforcing even if all human-caused greenhouse gas emission ceases. At this point, the

86 Romm, "How the EPA and New York Times Are Getting
     Methane All Wrong."

climate will propel the planet into a permanently hotter state. Most scientists seem to think that humans could still survive in this permanently hotter world, though they acknowledge that life would be considerably less pleasant. If we pass into a permanently hotter world *and* continue to burn all the fossil fuels we can, we could conceivably arrive at a "runaway greenhouse effect" in which the entire planet becomes uninhabitable for all life.[87]

So which time period makes the most sense for evaluating the warming impact of methane? The FAO favored the one-hundred-year time frame, LCC came down in favor of the twenty-year time frame, and this explains roughly 20 percent of the difference in their numbers.

LCC argued that climate carbon feedbacks have brought us close to tipping points such as those discussed above. If we don't act now (or in the next twenty years), we will find ourselves already beyond irreversible climate tipping points, and the further impacts of methane won't matter. In fact, the IPCC has warned that we have even *less* time than twenty years—until 2030—to limit climate catastrophe,[88] so perhaps we should be operating *within a ten-year time frame or less.*

It may already be too late to avoid dangerous climate feedbacks. In summer 2014, oilmen in Siberia on the Yamal Peninsula[89] discovered giant methane sinkholes, some several hundred feet wide and several hundred feet deep, that are releasing methane. Permafrost is also disintegrating in Alaska, Scandinavia, and Canada, where scientists fear large-scale releases of carbon dioxide[90] and presumably methane as well.

....

87 Hansen, *Storms of My Grandchildren*, p. 236.
88 Watts, "We have 12 years left to limit climate catastrophe, warns UN."
89 Staalesen, "New Sinkholes Appear in Yamal."
90 Berwyn, "Massive Permafrost Thaw Documented in Canada"; Kokelj, "Climate-driven thaw of permafrost preserved glacial landscapes."

The effects of climate change are now spinning out of control.

3. *Land use* (items 1 and 2 in Box 7-1) accounts for about 45 percent of the discrepancy between LCC and the FAO's earlier report. "By itself, leaving a significant amount of tropical land used for grazing livestock and growing feed to regenerate as forest could potentially mitigate *as much as half (or even more) of all anthropogenic GHGs* [greenhouse gas emissions]."[91] Elsewhere, Robert Goodland said that through land-use changes, we could bring atmospheric carbon back down to pre-industrial levels.[92]

These conclusions were tantalizing, but how did LCC arrive at them? Measuring livestock respiration accounted for the greatest portion of the land-use effects in question. Livestock respiration? This point is confusing to many. The authors later explained that "livestock respiration" could be understood as a proxy for land use. It approximates the natural carbon sequestration that *would* have occurred had we not grazed the land; the carbon in the CO_2 from livestock respiration is roughly equivalent to the carbon in the plants the cattle eat. The authors used this measure because they could not find a better way to quantify the CO_2 impacts of land use.[93]

---

91 Goodland and Anhang, "Livestock and Climate Change," p. 13 (emphasis in original).
92 Goodland, "How to Reverse Climate Change Before It's Too Late."
93 Goodland, "'Livestock and Climate Change': Critical Comments and Responses."

Category	Annual greenhouse gas emissions	Percentage of total
1. Livestock respiration	8,769	13.7%
2. Uncounted land use (could be much more)	2,672+	4.2%+
3. Uncounted methane	5,047	7.9%
4. Undercounted items (e.g., increase in livestock numbers, out-of-date data)	5,560+	8.7%+
5. Misallocated (e.g., waste, packaging, medical care)	3,000+	4.7%
Original FAO estimate (now less than 18 percent because of additional greenhouse gas emissions found)	7,516	11.8%
TOTAL FROM LIVESTOCK	**32,564**	**51.0%+**

Box 7-1. "Livestock and Climate Change" estimates

Source: Goodland and Anhang, "Livestock and Climate Change." Annual greenhouse gas emissions are in millions of tons of CO_2-equivalent.

Land use, forests, and carbon dioxide

Estimates of land-use impacts on climate are harder to conceptualize than estimates of the impacts of burning coal, which require only that we measure the *direct* addition of CO_2 into the atmosphere. Land-use impacts are *indirect* additions of CO_2: the carbon in the atmosphere that we are responsible for when our actions subvert natural biological processes. The concept of an "indirect emission" isn't complex, but it is a new concept for many people.

People often discuss *carbon sequestration* (pulling CO_2 out of the atmosphere and storing it somewhere) in terms of re-engineering coal-burning power plants to capture CO_2, compress it and inject it into rocks, and transport it to a secure storage site underground. Reforestation accomplishes the same thing, except that the storage site is a tree. It is *natural* carbon sequestration. Cattle grazing subverts reforestation and thus indirectly increases the CO_2 in the atmosphere.

We don't have to arrive at precise figures to see that the CO_2 effect of cattle grazing is significant, and it's probably a distraction to try to arrive at precise figures anyway. Grazing cattle *clearly* has a carbon impact when that land would otherwise return to forest. *Forests contain about three-quarters of the earth's plant biomass (phytomass), and about half of that is carbon.*[94] In Australia, forests contain about *ten times* as much carbon as grasslands per unit area.[95] Because of this extraordinary carbon-storage potential, many people argue that reforestation is a key strategy to deal with climate change.

Throughout history, humans have caused wide-ranging and extensive deforestation. As noted in chapter 5, without human land use, the roughly 450 gigatons of (sequestered) carbon in phytomass on the planet now would be 916 Gt C—*more than twice as much.*[96] For comparison, the total amount of carbon humans have directly *added* to the atmosphere is about 300 gigatons! In short, deforestation and reforestation can have a *profound* carbon impact. There have now been several new attempts to measure this impact:

....

94 O'Laughlin and Mahoney, "Forests and Carbon."
95 Australia's Chief Scientist, "Which plants store more carbon in Australia." "Australia's Chief Scientist" is a position in the Australian government (as of 2021, Dr. Cathy Foley).
96 Erb et al., "Unexpectedly large impact of forest management and grazing."

- One recent study found that even *excluding* existing forests, urban areas, or farm areas, we still have the opportunity to sequester about 200 Gt C.[97] By "farm areas," the authors meant *cropland*, not permanent pastureland or grassland. Presumably, this would allow for reforestation of much pastureland and grassland, with an even greater potential if we included reforestation of cropland currently producing livestock feed.

- Another recent study concluded, "Our analysis finds that consumption and LUCs [land use changes] can have many times larger implications for climate change than often calculated," by undervaluing alternatives "such as shifts to diets low in beef and milk."[98]

- Yet another study addressed plant-based diets, saying that excluding animal products "has transformative potential," reducing food's land use by about 3.1 billion hectares (a 76-percent reduction!), including a 19-percent reduction in arable land.[99]

- Another group of researchers said that a worldwide switch to plant-based diets would sequester an additional 332 to 547 Gt CO_2 (about 90 to 149 Gt C), which is equivalent to 99 to 163 percent of the CO_2 emissions budget consistent with a 66-percent chance of limiting warming to 1.5 degrees Celsius.[100]

- Sailesh Rao and his colleagues estimated that if the entire world allowed lands devoted to livestock to revert to their former state, enough land would return to forest to pull 265

---

97 Bastin et al., "The global tree restoration potential."
98 Searchinger et al., "Assessing the efficiency of changes in land use."
99 Poore and Nemicek, "Reducing food's environmental impacts."
100 Hayek et al., "The carbon opportunity cost of animal-sourced food production on land."

gigatons of carbon out of the atmosphere.[101]

When human activities disrupt natural biological processes to prevent absorption of CO_2 through photosynthesis, this should be counted as an *indirect* CO_2 emission. These indirect emissions of CO_2 aren't only harder to estimate, they're also harder to state in annual terms because forests grow at different rates. Most potential tree growth is found in relatively young forests.[102] Mature forests, by contrast, store a lot of carbon but may not be very good carbon sinks: they may have minimal capacity to absorb *additional* carbon. Still, even old-growth forests may be able to sequester carbon at the rate of six tons per hectare,[103] though in some arid regions, forests may not be the best way to absorb carbon because of their susceptibility to fires.[104]

"Carbon sequestration" can be accomplished by different ways: through reforestation, or through mechanically capturing CO_2, compressing it, and injecting it into rocks deep underground. But regardless of the method, the amount of carbon sequestered wouldn't necessarily be the same as the amount of carbon taken out of the atmosphere. Much of the CO_2 emitted by humans has been absorbed by the oceans, and the net result of sequestration would likely be to rebalance the carbon cycle, with lesser amounts of carbon in both the atmosphere and the oceans.

Scientific reticence about including land use in analysis of climate impacts is rapidly becoming a thing of the past. Reforesting land freed up from livestock agriculture would enable "the mass restoration of ecosystems and wildlife, pulling

101 Rao et al., "The Lifestyle Carbon Dividend."
102 Pugh, "Are young trees or old forests more important?"
103 Food and Agriculture Organization of the United Nations, "Forests and climate change."
104 Dass et al. "Grasslands may be more reliable carbon sinks than forests in California."

the living world back from the brink of ecological collapse and a sixth great extinction."[105] The IPCC outlines several pathways to limit damage from climate change; they all involve reforestation.[106]

It is beyond the scope of this book to render a precise judgment about the amounts and percentages of greenhouse gas emissions due to livestock agriculture. There are still a variety of estimates, and it is difficult to express the CO_2 emissions on a per-year basis. There does seem to be an emerging scientific consensus, however, that *the quantity of emissions due to livestock agriculture is very significant.* It is comparable to—perhaps greater than—the greenhouse gas emissions put directly into the atmosphere for all other purposes. We can't deal with climate change without dealing with livestock agriculture.

Conclusions

Climate change is much worse than we thought just a decade ago. In July 2016, David Carlson, director of the World Meteorological Organization (WMO), complained about "massive temperature hikes," increased flooding, and quicker ice-melt rates, saying, "What concerns me most is that we didn't anticipate these temperature jumps."[107]

A key reason we haven't accurately anticipated the current degree of climate change is likely that we haven't accounted for land use and livestock agriculture. Livestock agriculture was well underway (with some climate impacts) long before the Industrial Revolution, but the Industrial Revolution accelerated

....

105 Monbiot, "We can't keep eating as we are." Monbiot cites Searchinger et al., "Assessing the efficiency of changes in land use."
106 Watts, "We have 12 years left to limit climate catastrophe, warns UN."
107 Tabary, "Scientists caught off-guard."

and magnified those impacts. When the effect of livestock is added to the gigantic impact of just burning fossil fuels, we have a crisis of major proportions that puts fundamental limits on the growth of the human economy.

CHAPTER EIGHT

THE DIRT DILEMMA
The threat of soil erosion

Soil erosion is a slow-moving but inexorable threat. It will be a long time before, unchecked, soil erosion *by itself* will finally have measurable impacts on agricultural yields. However, unless we figure out how to synthesize food directly from inorganic materials, soil is indispensable.

Soil degradation can take many forms: not just soil erosion, but also salinization, nutrient depletion, acidification, soil compaction, loss of organic matter, and soil contamination.[108] We focus on soil erosion because erosion is easiest to quantify and understand; without soil, modern agriculture is impossible.

Soil erosion is destroying topsoil *much* faster than it is being formed. Soil degradation led to the collapse of past civilizations, most notably the ancient Mesopotamian civilizations in the Middle East and the Classic Maya Civilization in Central America in pre-European times. Unless we come up with a way

....

108 Koppitke et al., "Soil and the intensification of agriculture for global food security."

of dealing with soil erosion, it will eventually end our civilization as well.

Energy as a limit on agriculture

Amazingly, despite all this, our immediate problem isn't soil degradation but energy. Agriculture existed for thousands of years without fossil fuels or enormous energy use. Yet today, the single limiting factor on agriculture that is in *relatively* shortest supply isn't soil (or even irrigation water), but fossil fuels.

Through the Green Revolution, fossil fuels have allowed us to evade limitations associated with both soil and water. This is an example of "Liebig's law of the minimum" (see chapter 4). The single resource that is *relatively* the scarcest (among the many that are needed) is the limiting factor for the overall process. In this case, our immediate problem is oil rather than soil.

The Green Revolution has greatly increased soil productivity through use of artificial fertilizers, irrigation, and high-yielding hybrid varieties—all of which depend, to a lesser or greater degree, on energy. Since the Second World War, the Green Revolution has multiplied the productivity of agriculture several times over. Yields of corn harvested for grain in the United States are more than *six times greater* today than they were just a century ago.[109] Wheat yields tripled in the United States and rice yields nearly tripled in Japan.[110]

Without artificial fertilizers and related technology, we could support only about half of today's population even on overwhelmingly vegetarian diets, or only 40 percent of today's population on today's average diets.[111] We are in a "progress

....
109 United States Department of Agriculture, "Crop Production Historical Track Records," 29, 31. Corn harvested for grain yielded 26.2 bushels in 1917, but 176.6 bushels in 2017.
110 Smil (2011), "Nitrogen cycle and world food production."
111 Smil (2002). "Nitrogen and food production."

trap"[112]—we can't easily go back to the old technology of just a century ago and still feed everyone. Due to agricultural successes, human population has increased far beyond what nineteenth-century agriculture could support.

Soil erosion as a fundamental problem

No matter how much energy we have, though, we will *still* need fertile land to grow food. Since soil is continually being formed (though quite slowly) from decomposed rock, it is in principle a renewable resource. But unfortunately, soil erosion is proceeding at a much faster rate than the rate of natural soil formation—at least ten times greater, and in much of the world even faster than that. On top of that, the time frame for remedying soil erosion problems is on the order of *centuries*. In fact, depending on what we expect in terms of a "solution" to soil erosion, it may be "unsolvable" already.

Bringing additional land into production is at best only a temporary expedient to address soil erosion. Because of the "best first" principle,[113] we have already brought the best land into agriculture. Given several different ways of obtaining a needed resource, humans will pick the one that requires the least effort (the "low-hanging fruit"). Any new land brought into production today typically incurs various unpleasant costs we haven't seen previously, such as destroying what's left of the Amazon rainforest or draining groundwater supplies.

Soil erosion is, to a certain extent, an inevitable consequence of human cultivation of the land, when bare soil is exposed to wind and rain. Any vegetative cover at all over the soil will break the velocity with which rain strikes the surface, resulting in an

112 See Ronald Wright, *A Short History of Progress*, for more on progress traps.
113 Ricardo, *On the Principles of Political Economy and Taxation*.

enormous decrease in its kinetic energy. The formula for kinetic energy is:

$$KE = \tfrac{1}{2} mv^2 \text{ (where KE = kinetic energy, m = mass,}$$
$$\text{and v = velocity).}$$

The greater the kinetic energy, the greater the soil erosion. Decreasing the *velocity* of rain as it strikes the soil is the key here to controlling soil erosion. Obviously, you need rain to grow crops, so you don't want to decrease the *mass* of water. Yet by design, most farming will till the soil to remove weeds at the beginning of the year, leaving the soil bare for at least a short period of time.

Total soil eroded in the entire world annually is estimated to be 75 billion tons, and this is probably a conservative estimate. The relatively best annual soil-erosion rates are in the United States and Europe, where erosion is estimated at 10 tons per hectare per year (t/ha/yr) on cropland.[114] Worldwide, soil erosion on cropland averages 30 t/ha/yr, and in China it is 40 t/ha/yr.[115]

One hectare is about 2.47 acres. To give us an idea of what *tons per hectare per year means*, one inch of topsoil on a hectare of land would weigh about 350 to 400 tons—more or less, depending on the water content of the soil, the structure of the soil, and so forth. This means that, every thirty-five to forty years, we lose an inch of topsoil everywhere—and that's the most optimistic estimate. China, for example, might be losing an inch of topsoil each *decade*.

....

114 Pimentel and Pimentel, "Soil erosion," pp. 205, 206.
115 Pimentel and Pimentel, "World Population, Food, Natural Resources, and Survival," p. 151.

Soil formation

The good news is that soil erosion isn't a one-way street. Soil *formation* is also always going on, through the process of "weathering." Soil forms as part of this weathering of the parent material (rocks or rocky soil) that lies beneath the existing topsoil, due to the weight of the soil above it and the contact of the parent material with water trickling down from above. The bad news is that soil formation happens at time frames longer than we're used to thinking about. It can take many *centuries* to form one inch of topsoil, and we are eroding topsoil much faster than that.

There is general agreement about the rate of soil formation, with an impressive list of authorities putting it in the region of 0.5 to 1.0 t/ha/yr (see Box 8-1, "Measuring soil formation"). This is estimated as a rate of about one inch of topsoil every five hundred years.[116] Comparing the erosion rate of 10 t/ha/yr (the most optimistic estimate) to a rate of soil formation of 0.5 to 1.0 t/ha/yr leads us to the conclusion that, *even in the most favorable circumstances, soil is eroding ten to twenty times faster than it is being formed.* On pastureland and grazing land, the erosion is somewhat less, about 6 t/ha/yr,[117] which means that soil is "only" eroding at six to twelve times the rate at which it is being formed.

116 Pimentel and Pimentel, "World Population, Food, Natural
    Resources, and Survival," 151.
117 Pimentel and Pimentel, "Soil Erosion," p. 206.

Year	Source	Rate of soil formation
1909	T. C. Chamberlin, cited in Leonard C. Johnson, "Soil loss tolerance."	One inch of topsoil in 1000 years (about 0.35–0.4 tons / ha / year)
1939	H. H. Bennett (founder of the US Conservation Service), cited in Leonard C. Johnson, "Soil loss tolerance."	One inch of topsoil in 300–1000 years (about 0.35–1.3 tons / ha / year)
1998	Young, *Land Resources*.	One ton / hectare / year
2004	Rate that would be typical for geologic erosion from gently sloping soils. Troeh et al., *Soil and Water Conservation*, p. 69.	One ton / hectare / year
2004	Natural soil formation in central Texas, based on analysis of effects of cosmic rays on nuclides in soils. Troeh et al., *Soil and Water Conservation*, pp. 68–69.	0.18–1.8 tons / hectare / year
2005	Troeh and Thompson, *Soils and Soil Fertility*, p. 392.	About 0.2 ton of soil per acre each year (about 0.4 ton/ha)
2006	Pimentel, D., "Soil Erosion," p. 205.	0.5–1 ton / hectare / year

Box 8-1. Measuring soil formation

Most of these estimates work through the inference that *natural* soil formation must equal or exceed *natural* soil erosion (otherwise, there would be no topsoil at all after millions of years). One estimate is based on effects of cosmic rays on soil (to determine how long nuclides in the soil have been close enough to the surface to be affected by cosmic rays).

The rate of soil formation varies quite a bit depending on the condition of the soils. Generally speaking, the deeper the topsoil, the slower the soil formation taking place below the topsoil, and vice versa. Deeper topsoil means more "insulation" between the parent material and the weathering agents, thus slowing down the weathering and the process of soil formation. In natural conditions, "eventually soil depth will stabilize when the soil is thin enough to permit weathering that produces new soil as fast as the rate of soil erosion."[118]

118 Troeh and Thompson, *Soils and Soil Fertility*, p. 392.

The problem of cattle grazing

Most agricultural land in the world is grazing land for cattle or other ruminant animals that are capable of eating grass. In the United States, this is about 300 million hectares (750 million acres).[119] Grazing animals is an ancient practice. In human economic terms, it is quite efficient. It requires land, but not much human labor or energy; animals are turned loose in a pasture and feed themselves. Modern rangeland beef requires 5 percent of the energy required to produce the same number of calories of feedlot beef.[120] Soil erosion is somewhat less on grazing land than on cropland, about 6 t/ha/yr on pasture and grazing land in the United States. But this modest improvement in soil erosion rates is more than offset by the fact that the caloric yields per acre of grazing land is much, *much* less than that of cropland. The soil erosion *per calorie produced* from grazing animals is *on the order of one hundred times greater* than the soil erosion per calorie from growing crops.[121] Defenders of livestock agriculture sometimes argue that much grazing land couldn't be used for growing crops for people and therefore is "wasted" if

119 527 million acres of non-federal land was pastureland or rangeland in 2012: United States Department of Agriculture, *2012 National Resources Inventory*, p. 2-1. To which we can add more than 230 million acres of federal BLM land grazed and the federal forest land grazed in 2004: Glaser et al., *Costs and Consequences*, pp. 10, 11.
120 Steinhart and Steinhart, "Energy Use in the U.S. Food System," p. 312.
121 In Akers, *A Vegetarian Sourcebook*, p. 119, I calculated that the erosion from rangeland beef was six times greater than feedlot beef and more than 150 times greater than just eating the crops directly. This doubtless captures the order of magnitude. The exact figure, however, would be almost impossible to calculate, due to the huge variation in grazing intensities and crops that are grown. Moreover, even "feedlot" animals are usually grazed at the beginning of their short lives.

not used for cattle. This makes sense only if we are determined to use every last square inch of the planet for human purposes. The rather minimal food return from grazing causes tremendous ecological damage.

Historically, grazing has been the most environmentally destructive form of human activity. Overgrazing has led to soil erosion and desertification through cattle trampling, compacting, and pulverizing the soil, as well as by directly removing vegetation. Soil compaction from grazing makes the land susceptible to wind erosion, as the finer particles are more easily blown away. Soil compaction also makes it more difficult for rain to penetrate the soil; the water tends to run off and the soil underneath dries out.[122] Changes in land use can also alter the local rainfall in an area; deforestation sometimes decreases local precipitation.[123] Overgrazing thus has the potential not only to destroy soil and vegetation, but also to create deserts.

Soil erosion in history

If soil erosion is this bad—soil eroding at ten to eighty times faster than it is being formed—then why haven't we heard about this already? Why hasn't it *already* resulted in some sort of ecological catastrophe? After all, agriculture has been practiced for ten thousand years now.

Part of the answer is that soil erosion is a slow-moving process and that agriculture throughout most of history was only a small fraction of what it is today. Human population was less than 1 percent of today's population just three thousand years

122 See Warren and Maizels, "Ecological Change and Desertification"; Novikoff, "Traditional Grazing Practices and Their Adaptation to Modern Conditions"; and Lundholm, "Domestic Animals in Arid Ecosystems."
123 Pielke et al., "An overview of regional land-use and landcover impacts on rainfall."

ago, and less than 10 percent about three hundred years ago. But the other part of the answer is that even with a much smaller human population, ecological catastrophes *have* occurred. The land destroyed by human activities (soil erosion, salinization of soils, overgrazing, and so forth) is 20 million square kilometers. This is about 15 percent of the earth's land area—*an amount that exceeds the total amount of cropland currently in cultivation.*[124]

When the Europeans first arrived in America, they found topsoil that was rich and widespread because the indigenous peoples didn't have the same large-scale agricultural enterprises that the Europeans did. North America, even today, has some of the best agricultural lands in the world. But American agricultural methods have reduced soil resources by at least half; "a systematic if unconscious rape of the land has had an impact that rivals or exceeds that of 6 to 10 millennia of cultivation in the Mediterranean world."[125]

Ways of dealing with soil erosion

Soil erosion hasn't attracted nearly as much attention as other environmental issues, and approaches to soil erosion are not very well discussed. But here are some leading possibilities for dealing with soil erosion.

1. *Fertilizers.* Applying artificial fertilizers (especially nitrogen fertilizers) is the default approach to declining or inadequate soil quality. Nutrient-poor soil? Just add fertilizers. However, we cannot rely indefinitely on fertilizers to bail us out.

Artificial nitrogen fertilizers are currently produced using natural gas, a fossil fuel, which itself is finite and depleting. Even if produced renewably from ammonia using non-carbon sources (solar, wind, or nuclear), they would still require energy from somewhere. In many areas, applying more artificial fertilizers

124 Kovda, "Soil Reclamation and Food Production," pp. 160–161.
125 Butzer, "Accelerated Soil Erosion."

doesn't really increase crop productivity, and overfertilization can create soil acidification.[126] Nitrogen fertilizers can be synthesized, taking abundant nitrogen from the air, but plants also need numerous other nutrients which must be mined, and these supplies are obviously finite.

Especially troubling is the situation regarding phosphorus, a key plant nutrient. Much of the phosphorus added to the land in fertilizers currently winds up in the oceans. In the past, wild animals played a key role in returning phosphorus from the ocean to the land areas, through their carcasses, urine, and dung. But by destroying wildlife (see chapter 6), we have disrupted these natural systems and reduced this natural phosphorus supply by more than 90 percent. Already there is concern about the "phosphorus crisis" and "peak phosphorus."[127]

2. *Land reform.* There are several political approaches that might be effective. High contemporary rates of soil erosion *may* be due to the widespread use of industrial monocultures. We could break up large monocultures into smaller varied allotments cared for by better motivated and better trained small-scale farmers. It may also be that 80 percent of the soil erosion comes from degradation of 20 percent of the land, which could very well be the most erodible land, only farmed out of desperation by the poor because no better land is available. Providing social justice for the poor (such as a universal basic income, discussed in chapter 21) and taking this highly erodible land out of production might eliminate 80 percent of the problem immediately.

3. *No-till agriculture* is a technical solution that eliminates tillage altogether. However, this is typically an energy-intensive

---

126 Koppitke et al., "Soil and the intensification of agriculture for global food security."
127 Kean, "Peak phosporus?"; Swenson, "What Can Be Done About the Phosphorus Crisis?"; Abraham et al., "The Sixth R."

and capital-intensive process that uses chemical pesticides and lots of heavy machinery to control weeds in the absence of tillage. (Tillage is the "traditional" tool to control weeds). With genetically modified organisms (GMOs), some farmers using no-till techniques have simply planted crops that have been modified to be resistant to their pesticides then blanketed their entire fields in pesticide. This practice can't be good for the soil in the long run; it is the diametric opposite of "organic" approaches.[128] One prominent geologist, however, has advocated *organic* no-till agriculture,[129] an approach that deserves further investigation.

4. *Organic agriculture or veganic agriculture.* Organic agriculture (as defined in the United States) dispenses with artificial fertilizers and pesticides, using livestock manure as fertilizer; "veganic agriculture" (organic agriculture without livestock) would use plants directly as manures. There is concern about whether such approaches could produce as much food as we do today.[130] Both organic and veganic agriculture would seem to require more land than using artificial fertilizers, either to support the livestock that provides the manures, or to grow the green manures (veganic). So, even though per-acre yields might be just as good or even better, the total land required might increase. Moreover, while they might be better systems in other ways, in terms of *soil erosion*, it's not clear how organic or veganic agriculture is an improvement. Growers would probably still till the soil and we could still have industrially produced organic or even veganic monocultures, allowing for soil erosion to set in.

....

128 See discussion in chapter 13 of Carlisle, *Lentil Underground.*
129 Montgomery, *Growing a Revolution.*
130 Two pro and con articles: (1) Badgley et al., "Organic agriculture and the global food supply," and (2) Connor, "Organic agriculture cannot feed the world."

5. *Permaculture* is a school of thought that emphasizes agriculture that resembles natural ecosystems or exists in a symbiotic relationship with them. It isn't uniformly defined, which makes scientific analysis more difficult. But such practices as crop rotations, intercropping, cover crops, and minimum tillage[131] may minimize soil erosion and save the day. Permaculture theory appears to address soil erosion by minimizing tillage and keeping ground cover as much as possible. But how much land would this require? How labor-intensive would it be? What would yields be? How would such techniques be taught and spread worldwide? And most obviously, how much would soil erosion be reduced? Permaculture might be a *better* system but still not be able to sustainably feed eight billion humans. We would need to answer all these questions (and perhaps others) to really understand the role of permaculture.

6. *Vegan diets and/or population reduction.* We could decrease the total amount of food required from the agricultural system through population reduction, plant-based diets, or (most likely) both. Population reduction means there are fewer people to feed; plant-based diets mean less land is required per person. By decreasing agricultural intensity, we could simply leave agricultural land fallow—left unplanted for a time—to allow the soil to regenerate.

How would this work? As a preliminary ballpark estimate, if soil were eroding "only" ten times faster than it was being formed, we could leave 90 percent of agricultural land fallow, allowing it to recover so that some would be available when the existing agricultural land was too exhausted to continue growing food. Even with universal veganism, in the long run (over centuries) it may not be possible to feed the current world population on just 10 percent of today's agricultural land with

131 Food and Agriculture Organization of the United Nations, "What are the environmental benefits of organic agriculture?"

today's rates of soil erosion; we would likely also need to reduce human population.

7. *Miscellaneous technical approaches.* Some have suggested the use of biochar, a kind of charcoal produced by burning biomass in a low-oxygen environment. The idea is that such use would accelerate soil formation; but so far, claims for biochar haven't been documented. Hydroponic agriculture eliminates the use of land altogether, but the costs in terms of energy and dollars are about ten times that of conventional agriculture, clearly not sustainable on any sort of mass scale.[132]

There are plenty of unknowns here. What we really need is a set of techniques that won't only largely avoid soil erosion, but also avoid nutrient depletion and maintain Green Revolution yields. New techniques should be able to disseminate throughout the world in the absence of any clear economic reasons to do so. Currently, it seems that the very Green Revolution techniques that have increased yields dramatically often also have the effect of degrading soil. While it is tiresome to say and to hear, we need further research on this subject, which is so clearly of interest to the future of humanity. In the meantime, our farming practices continue to be intensely unsustainable with no obvious solutions in sight.

Conclusions

Soil erosion is a mammoth, slow-moving, and intractable problem. Soil erosion forces us to question whether we can feed the world even on a diet exclusively of plant foods. In truth, our *immediate* food problems (in the next decade or so) are more likely to come through a new energy crisis or social disorder that will make high-yielding industrial agriculture harder to

132 Pimentel and Pimentel, "World Population, Food, Natural Resources, and Survival," p. 151.

maintain. But over the course of centuries, soil erosion could be fatal even for a "post-industrial" low-energy civilization. There may be solutions, but until we clearly know what they are, soil constitutes a major limit for the human economy.

CHAPTER NINE

WATER AND AGRICULTURE
Draining out the source of life

Water is an absolute limit on the growth of the food economy. While water is used for many different economic purposes—drinking, washing, and industrial processes—agriculture is the overwhelmingly predominant use. Worldwide, agriculture accounts for about two-thirds of all water use and more than 90 percent of all water consumption.[133] (*Consumption* is a technical term, explained below, that refers to water removed from the rest of the hydrological cycle.) Experts refer to water for agriculture as either *green water* (water falling from the sky) or *blue water* (water diverted from rivers, lakes, etc., for irrigation).

There is good news and bad news about water. The good news is that we're not going to run out of water. Thanks to the hydrological cycle, water is constantly re-entering our agricultural lands through rain (the green water).

But today, we have a triple dose of bad news about water. For millennia, rainwater (green water) provided all the water that most people needed. But, thanks to climate change, droughts are

---

133 Richter, *Chasing Water*, p. 25, Table 2-1.

increasing and green water supplies are becoming more erratic. More bad news: even in the best of circumstances, rainwater for croplands isn't enough. To maintain our agricultural output, we need irrigation water from streams, rivers, lakes, or groundwater—the blue water that doesn't come directly from rainwater. The third dose of bad news is that blue water is not only limited; it is depleting.

Water technology has put us in another "progress trap." Irrigated cropland is *much* more productive than non-irrigated cropland,[134] and is a key component of the Green Revolution. Irrigation has expanded fivefold since 1900,[135] and 2.5 million square kilometers of cropland is irrigated in some fashion.[136] Irrigated land is about 17 percent of the total 15 million square kilometers of all cropland. But this small portion of irrigated land provides almost half the total value of crop production.[137] There is no practical way we can dispense with irrigation and maintain anything close to our agricultural output.

Problems with irrigation

Agriculture doesn't just use water, it *consumes* water. Water can be used but *not* consumed: it is the water used taking a shower, cleaning the dishes, or cooling nuclear reactors. Such water continues downstream even after being used. Sometimes it is polluted or dirty, but in principle (and many times, in practice as well) it can be cleaned and re-used for another purpose.

Consumed water, however, is water that disappears from the rest of its downstream journey. How does this happen? There

....

134 Kucharik and Ramankutty, "Trends and Variability in U.S. Corn Yields."

135 Rosegrant et al., *World Water and Food to 2025*, p. 1.

136 Rosegrant et al., *World Water and Food to 2025*, p. 1.

137 Shiklomonov, *World Water Resources*, p. 19. And this estimate is from 1998, so it could be higher today.

are several ways. Most obviously, it can be evaporated into the atmosphere. Eventually, evaporated water will reappear as rain somewhere else, but it will not make it to the ocean or any further downstream. It can also become incorporated into a plant or a tree, where it will stay until the plant dies (or the plant "respires" water back into the atmosphere). This water also won't make it any further downstream.

The Colorado River runs dry before it reaches the Pacific Ocean because of water consumption. There's plenty of irrigation going on in the Colorado River Basin, much of it for growing animal feed, and this water is incorporated into the plants or is lost in evaporation or plant transpiration. Nearly half of all agricultural water is consumed,[138] and it will never make it any further downstream. Although water in a region is *renewable*, it is also *finite*. There is only so much water in the Colorado River Basin, and when it's all gone, it's gone until the next time it rains or snows upstream.

Because so much blue water is consumed, agricultural water use can have a major impact on the hydrology of a region. One well-known water disaster concerns the fate of the Aral Sea in Asia on the border between Kazakhstan and Uzbekistan. The rivers supplying the Aral Sea were so heavily used for irrigation that the Aral Sea has mostly disappeared. A similar problem affects the Dead Sea in Israel, which is also steadily shrinking because of water consumption by agriculture. In the Colorado River Basin in the western United States, water is nearly 100 percent consumed at various points along the way; the river entirely disappears by the time it gets to the river's delta in Mexico, at the Gulf of California. Irrigated agriculture is the main culprit.

Yet people continue to move into this area; as of 2016, Colorado was one of the fastest-growing states in the country.

138 Reig, "What's the Difference Between Water Use and Water Consumption?"

Las Vegas, Nevada, is also growing rapidly, and Phoenix, Arizona, is the fastest-growing city in the nation. Both cities are also in the Colorado River Basin and built in the middle of deserts.

These kinds of problems have led to epic and expensive water projects to provide water from elsewhere—perhaps by pumping in water from a more water-abundant region, or desalinating seawater. In July 2022, Arizona Governor Doug Ducey signed into law a bill which will spend one billion dollars on water projects, including building desalination plants in Mexico.[139] (Arizona would then take some of Mexico's water allotment from the Colorado River.) This doesn't solve the problem of water shortages; it just displaces it from a problem of water to of a problem of whatever resources are needed (concrete, energy, etc.) to move the water where we want it. No one wants to acknowledge the obvious: water is already a limit on our agricultural system.

Groundwater depletion

Where do we get irrigation water? Streamflow—as long as you have the energy, materials, and fortitude to pump it where it is needed—is as renewable as the overall precipitation that fills rivers and streams (although these days even precipitation is increasingly compromised by climate change). Groundwater, though, is limited in a more fundamental way. Groundwater can be depleted, just like fossil fuels. While some groundwater supplies can be recharged from fresh rainwater or wastewater, most cannot be. Groundwater is only 6 percent renewable, according to University of Victoria hydrologist Tom Gleeson

----

139 Loomis, "Pipelines? Desalination? Turf removal?", and Office of the Governor Doug Ducey, July 6, 2022.

and his colleagues in a widely cited 2015 article in *Nature Geoscience*.[140]

Global groundwater use accounts for about 20 percent of total irrigation water use: from 750 to 800 cubic kilometers (km³)[141] annually, compared to all water withdrawals, which are on the order of 3800 km³ per year. (To give you a more concrete idea, 1 km³ = 1,000,000,000,000 liters—that's one trillion liters in English.) However, in areas that depend heavily on groundwater, the effect of groundwater depletion is bound to be devastating. And three of the countries that *do* rely heavily on groundwater are the United States, India, and China. Oops!

The United States is one of the nations most dependent on groundwater for agriculture. Aquifers provide 65 percent of US irrigation water.[142] Two prominent agricultural regions critically depend on groundwater: the High Plains region, under which resides the Ogallala Aquifer, and the California basin, the source of perhaps 80 percent of the country's fruits, nuts, and vegetables. Groundwater depletion in the areas that rely on groundwater means that, in these areas, eventually *agriculture as we know it could simply cease*.

As the water table falls due to groundwater depletion, water-occupied pore spaces and cracks in the ground empty out and are often crushed by the weight of the land above them. This results in land *subsidence*: the level of the ground itself falls because of groundwater depletion. This is happening in the San Joaquin Valley, home of much of California's agriculture, where a famous photograph shows the collapse of the land from 1925 to 1977 (see box 9-1, "Land subsidence in the San Joaquin Valley"). Farmers were able to alleviate this problem for a while by using surface water rather than groundwater. But sometimes, due to

140 Gleeson et al., "The global volume and distribution of modern groundwater."
141 Shah et al., "The Global Groundwater Situation."
142 Pimentel et al., "Water Resources," p. 185.

drought or other demands on surface water, the surface water hasn't been sufficient, so groundwater pumping began again. Due to land subsidence, most areas are sinking about an inch every year, but some areas are sinking as much as an inch every *month.*[143]

Box 9-1. Land subsidence in the San Joaquin Valley

Source: United States Geological Survey, "Land Subsidence in the United States." This photo shows Dr. Joseph Poland standing beside a telephone pole in the San Joaquin Valley in 1977, illustrating how much the land's surface has declined since the underground aquifer was first pumped to irrigate fruits and vegetables. Since 1977, the land has subsided even further.

143 Schmit, "In California, Demand for Groundwater Causing Huge Swaths of Land to Sink."

The Ogallala Aquifer is a gigantic underground aquifer underlying much of the Great Plains, stretching from southern South Dakota to the Texas panhandle.[144] Experts have warned for decades that the Ogallala has become increasingly depleted, but we have done nothing except periodically wring our hands. The depletion varies widely from place to place: in Nebraska, there has been almost no depletion, but in Texas, there are counties where irrigation based on groundwater has already become impossible or cost-prohibitive. Other counties have less than fifteen years' worth of usable groundwater left, and the water table has dropped almost a foot every year for at least the past four decades.[145]

The disastrous Dust Bowl of the 1930s occurred in just this area of the Great Plains on top of the Ogallala Aquifer. (The existence of the Ogallala Aquifer was unknown at that time.) A key reason for the Dust Bowl was that in the 1920s, farmers had plowed up millions of acres of deep-rooted prairie grasses. When a drought arrived in the 1930s, the crops planted couldn't thrive and the soil blew away. Our agriculture hasn't changed that much since the time of the Dust Bowl; we are simply draining the aquifer to hold back the desert. Many farmers in the Great Plains have a real fear that the Dust Bowl will return.[146]

The United States isn't alone in this problem. It is much worse in India, where fully 60 percent of all grain production depends on groundwater.[147] India is the most groundwater-dependent country in the world, withdrawing 210 km^3 annually, almost

---

144 It includes parts of the states of Wyoming, South Dakota, Nebraska, Colorado, Kansas, Oklahoma, New Mexico, and Texas.
145 Brambila, "Ogallala Aquifer's dramatic drying" and Scanlon et al., "Groundwater depletion and sustainability of irrigation."
146 Laurence, "US farmers fear the return of the Dust Bowl."
147 Shah et al., "The Global Groundwater Situation," p. 6.

twice as much as the United States.[148] The price tag of India's intensive groundwater development, though, is declining water tables and saltwater intrusion into aquifers and soils.[149] Sooner or later, as groundwater supplies dwindle, it would seem that India will have to import many of its crops as well as deal with unemployed farmers: more than half the population is employed by farming.

The Green Revolution can take credit both for the vastly increased yields and the resulting resource problems due to groundwater depletion. It is a classic case of *shifting* resource limits rather than dealing with them. Groundwater depletion won't bring agriculture *everywhere* to a halt, but it will stop or severely limit agriculture in affected areas, as is already happening in parts of the world.

Salinization of soils

Irrigation also tends to lead to *soil salinization*, the buildup of salts in the soil to toxic levels, which is already affecting India among many other places around the world dependent on irrigation. Unlike rainwater, stream water and groundwater contain tiny but significant amounts of minerals and salts dissolved in the water.

These minerals aren't significant in tap water, where they are scarcely noticeable. But they *are* significant in irrigation water when that water is used for many years. When irrigation water is used, about half of it will evaporate back into the atmosphere, and when the water evaporates, it leaves the minerals behind.

....

148 Villholth, "Comprehensive Assessment of Consequences," 7. By comparison, annual groundwater withdrawals are 107 km³ for the US, 75 km³ for China, and 60 km³ for Pakistan, out of 750–800 km³ for the entire world.

149 Villholth, "Comprehensive Assessment of Consequences," pp. 2–3.

Thus, over a period of years, decades, or centuries, these minerals tend to build up in the soil. The ultimate result is soil salinization.

Soil salinization led to the collapse of the ancient Sumerian civilizations in Mesopotamia, site of one of the world's first experiments with widescale irrigation. Over hundreds of years, salts and minerals gradually built up in their fields, and the soils became saltier and saltier. For a while, the Sumerians were able to grow barley instead of wheat, because barley is more salt-tolerant. But eventually, even barley couldn't grow, and the area was abandoned as its population plummeted. The desertification of the "Fertile Crescent," as the region became known, was one of the first major ecological disasters in human history.

Today, salinization affects half of all irrigated soils, destroying 100,000 km^2 each year.[150] At least the people in ancient Mesopotamia could migrate to other areas; today, our agricultural system straddles the entire planet, and migration to other planets doesn't seem like a viable alternative.

Conclusions

Limits on water mean limits on agriculture. Irrigated land is extremely productive and irrigation is indispensable for our ample agricultural output. It is an indispensable facet of the Green Revolution that so generously increased food yields.

Not only is there not enough water, but we are losing water from the cycle through groundwater depletion, which especially affects the United States and India. We are also losing land to soil salinization. Water problems have ended civilizations in the past. Eventually, they could end ours as well.

150 Pimentel et al., "Water Resources," p. 189. 10 million hectares is 100,000 km^2.

CHAPTER TEN

PEAK OIL

Energy and the fate of industrial civilization

Peak oil, one of the iconic limits-to-growth issues, refers to the time when the world reaches the maximum point of oil production, after which oil production begins to decline, never to rise again to that level. Because so much in the world economy relies on oil, peak oil will have enormous consequences. Oil price spikes, due to oil shortages, preceded most of the recessions that happened after the Second World War.[151]

When we use the term without any qualification, *peak oil* refers to the *near-term* maximum point of oil production for the *entire world*. There is controversy not only about when peak oil will occur, but whether it will even be such a bad thing. "The stone age did not end for lack of stones, and the oil age will not end for lack of oil," goes a saying attributed to Saudi Arabian Sheik Yamani. And of course, many climate advocates believe we should deliberately bring about peak oil by limiting greenhouse gas emissions.

....

151 Hamilton, "Historical Oil Shocks."

The peak oil debate

The American geologist M. King Hubbert (1903–1989) initiated the first phase of the modern debates about peak oil. Hubbert famously predicted in 1956 that US oil production would peak in 1970.[152] Indeed, United States oil production *did* reach record levels in 1970, levels that weren't exceeded for more than forty years—though the 1970 peak eventually was surpassed as a consequence of the so-called "shale revolution" beginning around 2008. Following Hubbert's death in 1989, petroleum professionals Colin Campbell and Jean Laherrère initiated the second phase of the peak oil debate with a 1998 *Scientific American* article titled "The End of Cheap Oil." They cautiously predicted an end to *cheap* oil, saying that *conventional* oil would peak early in the twenty-first century[153]—a prediction that has turned out to be fairly accurate, at least in terms of rising prices.[154]

Definitions of *conventional* and *unconventional* oil (discussed below) vary. The important thing to know at this point is that conventional oil is easier to get out of the ground. There is no uniform agreement on the exact date, but conventional oil probably peaked in 2008. We can never be certain of the date of the peak; even years after a decline starts, there is theoretically the possibility of some new mind-bending oil discovery. But with discoveries declining year after year, despite geologists combing the globe looking for new deposits, that possibility becomes less and less likely every year.

....

152 Hubbert, "Nuclear Energy and the Fossil Fuels."
153 Campbell and Laherrère, "The End of Cheap Oil."
154 Bardi, "Peak oil, 20 years later." Whether and when conventional oil peaked is nebulous because of the uncertainty of both the definition of "conventional oil" and the difficulty of getting accurate data about it.

Thanks to unconventional oil, we are nowhere near running out of oil in any literal sense. Total world oil production reached an inflection point around 2005; oil became harder to produce, and some thought that this was in fact the peak of world oil production.[155] But oil production continued to increase at a much slower pace, with the increases coming almost entirely from unconventional oil. From 2004 to 2017, oil production increased at less than 1 percent per year,[156] and the cheery forecasts in 2005 of a "large, unprecedented buildup of oil supply in the next few years"[157] turned out to be spectacularly inaccurate.

Many assume that the rise of such unconventional oil production implies that we don't have to worry about peak oil, or that climate change is a much more pressing problem than peak oil. While in the long run climate change is the more serious issue, the economic effects of limits on oil are here *today* and can only grow greater with time—and we may not even be at the peak yet.

Oil isn't only the leading energy source in the United States; it's also at the bottom of the economic chain for manufacture or distribution of countless other durable goods. It is easy to use but hard to replace. If we have problems with oil, we will likely have problems with *everything* in the economy.

Oil shortages have already produced several recessions. We discovered this in 1973 when OPEC (the Organization of the Petroleum Exporting Countries) embargoed oil sales to the United States (to protest US pro-Israeli policies), triggering a spike in oil prices and a recession. In 1979, the Iranian Revolution and subsequent Iran–Iraq war resulted in oil shortages, a rise in oil prices, then another recession. In 2008, there was another unexpected rise in oil prices—this time just due to the inability

155 Most notably Kenneth Deffeyes. See Deffeyes, *Beyond Oil.*
156 Statistica, "Global oil production."
157 Yergin, "It's not the end of the oil age."

116

of the industry to pump enough oil—occurring at the same time as the Great Recession of 2007–2009. However, it wasn't the peak of the total world supply of oil. The country started pumping unconventional oil, oil prices oscillated wildly, and supply crept slowly upward until 2018, which represents the current oil-production peak.

There's no doubt that oil extraction has become increasingly difficult in the twenty-first century. But when, if ever, will oil peak? Is it possible that, given the economic crash following the COVID-19 pandemic, we have already passed the peak of world oil production, and that 2018 was the year of peak oil production for the entire world?[158]

The rise of unconventional oil

As oil becomes progressively difficult to extract, the world has come to increasingly rely on unconventional oil.

Conventional oil is oil that comes out of the ground with standard drilling techniques and requires minimal refining and processing. Unconventional oil is the same as conventional oil by the time it arrives in your gas tank; *how* it arrives there makes it unconventional. Though there is no precise definition of unconventional oil, by general agreement the term encompasses a variety of new and more expensive techniques for oil extraction (see Box 10-1, "Unconventional oil").

---

158 Patterson, "Was 2018 the Peak?"

Box 10-1. Unconventional oil

Some categories of oil often described as "unconventional":

1. *Biofuels.* Fuel derived from plants, such as corn or sugarcane. In the US, this is mostly corn ethanol, using about one quarter of all land devoted to corn.

2. *Oil shale* (actually, the energy source is the kerogen inside the oil shale rocks). Never commercially produced.

3. *Shale oil*, also known as tight oil. This is oil recovered through fracking.

4. *Gas to liquids.* Theoretically possible but never commercially produced.

5. *Coal to liquids.* Used by the Germans in the Second World War and by South Africa under apartheid.

6. *Enhanced oil recovery.* A generic term for getting more oil out of "uneconomical" oil wells.

7. *Natural gas liquids.* Not oil at all. Byproducts of drilling for natural gas, such as butane, ethane, propane, isobutane, and pentane. Used for plastics, tires, home heating, or as a gasoline additive.

8. *Deepwater oil.* Oil extracted by drilling underneath the ocean as much as a mile deep. Examples are North Sea oil and Gulf of Mexico oil.

9. *Polar oil.* Deepwater oil extracted in extremely cold conditions near the poles. Never commercially produced.

10. *Heavy oil*, such as bitumen, oil sands, and tar sands. Very viscous (thick); requires extensive processing. Examples: oil from the Alberta tar sands and heavy oil from Venezuela.

The most politically contentious of these fuels are biofuels, shale oil, heavy oil, and deepwater oil. Biofuels are theoretically renewable, but as they are manufactured in the United States—in the form of corn ethanol—they take up an inordinate amount of agricultural land and contribute to climate change because so much fossil fuel is used to grown the corn. Deep-sea oil drilling is also more expensive and has resulted in one spectacular oil spill, the *Deepwater Horizon* oil spill in 2010, which resulted from the explosion of a British Petroleum Company (BP) drilling rig in the Gulf of Mexico.

Shale oil, the basis for most of the recent oil boom in the United States since the Great Recession, is oil obtained through hydraulic fracturing (or fracking), which involves injecting chemicals, sand, and water into rocks containing oil to access and extract the oil. Fracking is quite a bit more expensive than drilling for conventional oil, and oil companies have had great difficulty turning a profit on their investments; it has usually been a money-losing proposition, even with higher oil prices. Fracking sometimes also has contaminated groundwater and provoked anger when oil operations get too close to residential neighborhoods. Two fatal explosions in Colorado, one in a residential neighborhood, were traced to a fracking operation.

The most notorious example of heavy oil is the enormous and environmentally destructive operation in an underground oil reservoir in Alberta, Canada, known as the Alberta tar sands. The tar sands lie underneath more than fifty thousand square miles of boreal forest in northern Alberta, about the size of the state of Arkansas. The oil itself is thick and is mixed with sand, earth materials, and water; the oil first has to be separated from everything else, then refined to be less viscous, with the waste products strewn everywhere. The environmental damage has been immense; "The tar pits have slowly turned northern

Alberta into a giant toxic dump site."[159]

These technologies have expanded the supply of oil, but at a higher price: an economic price, an environmental price, and a political price. They are "new" only in the sense that their large-scale production is relatively recent. Both fracking and offshore oil drilling in some form are more than a century old. But deepwater drilling (in depths greater than one thousand feet) was quite limited before the mid-1990s,[160] and large-scale commercial hydraulic fracturing only began when oil prices started to rise sharply[161] (around 2005 to 2008). *Only recently has the price of oil risen high enough to justify the use of these technologies.*

The resource pyramid

Geology makes peak oil a simple issue. But economics and politics make peak oil a complex issue. Peak oil is simple because fossil fuels are limited and increasingly difficult to find and extract. Peak oil is complex because of the propensity of governments and businesses to use debt to postpone any financial day of reckoning, counting on future economic growth to pay it all back. It becomes doubly complex with new expensive unconventional oil technologies that have increased oil production.

We can understand this by looking at the idea of the resource pyramid (see Box 10-2, "The resource pyramid"). The resource pyramid is essentially a consequence of the "best first" principle: we look for the easiest resources first. But after we go through the easy resources, further exploitation becomes more and more difficult. We will likely never "run out" of oil, but at a certain

159 Moore, "Basins: The Alberta Tar Sands."
160 National Commission on the BP Deepwater Horizon Oil Spill
    and Offshore Drilling, "A Brief History of Offshore Oil Drilling."
161 Manfreda, "The Real History of Fracking."

level of difficulty in extracting the resource, people will look for alternatives or cut back their consumption. The "resource pyramid" idea applies to oil, but it also applies to other fossil fuels and to mineral resources as well. We will naturally go after the "easy gold" and "easy copper" first. Minerals will gradually become more difficult—but not impossible—to mine and process.

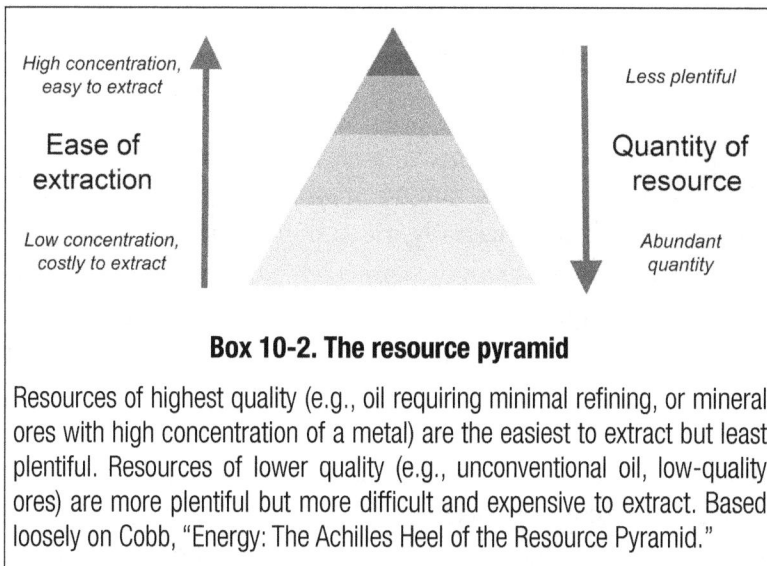

High concentration, easy to extract

Ease of extraction

Low concentration, costly to extract

Less plentiful

Quantity of resource

Abundant quantity

Box 10-2. The resource pyramid

Resources of highest quality (e.g., oil requiring minimal refining, or mineral ores with high concentration of a metal) are the easiest to extract but least plentiful. Resources of lower quality (e.g., unconventional oil, low-quality ores) are more plentiful but more difficult and expensive to extract. Based loosely on Cobb, "Energy: The Achilles Heel of the Resource Pyramid."

But increasing difficulty in resource extraction may manifest in ways other than an increase in *economic* price. Oil extraction may also impose a *political* price (as subsidies are marshaled to provide the oil) or an *environmental* price (as we are driven to increasingly destructive methods of extraction). Why would we even undertake such schemes as expensive, complex, polluting, and politically unpopular as fracking or the Alberta tar sands? Or drilling in the middle of the Gulf of Mexico underneath thousands of feet of water?

All these techniques are more expensive than conventional oil drilling, which is why they are only now being deployed. They are also environmentally destructive: fracking involves injecting polluting chemicals into the ground which sometimes get into groundwater; extracting oil from the Alberta tar sands uses vast quantities of water and has destroyed much of the boreal forest in Canada; and deep-sea drilling in the Gulf of Mexico resulted in the infamous *Deepwater Horizon* oil spill. Environmentalists have targeted all of them. We would only even *attempt* such things in the first place if we had nowhere else to turn. "Easy oil" is gone or in steep production decline.

The price of oil

Oil depletion is relentless and never sleeps. However, increasing scarcity doesn't automatically mean higher prices. To increase production of increasingly scarce oil, the price of oil needs to be *low* enough for consumers to afford, but *high* enough for producers to make a profit.[162] If *both* conditions are not fulfilled, something in the economy will not work: consumers will be poorer, or businesses will suffer bankruptcy, or both. The ultimate outcome of oil scarcity may be *lower* oil prices if the whole economy declines, as happened during the Great Recession, and as happened early in 2020 amid the COVID-19 pandemic. The same principle applies to other fossil fuels and natural resources, which may oscillate up or down based on a number of things unrelated to absolute scarcity.

Governments have tried various strategies to avoid such dire consequences. In an extreme case, the government could seize oil companies within its borders and operate them directly. This doesn't end the problem of costs; it just shifts the problem from an *economic* one to a *political* one. As long as the government can stay in power and stay solvent, any peak oil crisis can be postponed by lesser or greater government intervention.

....

162 Tverberg, "How the Economy Works as It Reaches Energy Limits."

To secure American access to Middle East oil, Bush tried invading Iraq in 2003, but that had unanticipated political consequences. The "quantitative easing" policy under the Obama administration (2009–2014) made it easier for everyone to borrow money, temporarily alleviating the problem of high oil prices through debt, and economic activity slowly recovered. The Trump administration tried to alleviate costs by eliminating pesky environmental regulations and opening new public lands to drilling. It is also possible for unwitting private investors to subsidize oil production; abundant hype about the shale revolution (fracking) has pushed money into oil companies, but these investments haven't yet produced profits.

Thus, the question of how intrinsic limits on oil will become manifest is a complex one. Since 2005, the price of oil has oscillated wildly, from less than 30 dollars a barrel to more than 147 dollars a barrel and everywhere in between. It even briefly dropped below zero in April 2020! This complex interaction of politics, economics, and technology that occurs when a resource becomes harder to extract is happening now and has been since about 2005.

Oil in relationship to other energy sources and mineral resources

The United States overwhelmingly depends on fossil fuels for energy, and all the fossil fuels are finite and rapidly depleting. There are additional supplies, but—like oil—they have also become harder to extract. However, we have omitted any lengthy discussion of "peak coal" or "peak natural gas." In practical terms, oil is a stand-in for *all* fossil fuels: when it declines, the whole energy system will likely decline.

If we put aside that it's not renewable and causes climate change and other environmental problems, oil would be an almost ideal fuel. It is widely used, compact, and easily transportable; it fits in perfectly with our vast infrastructure, with oil companies, roads, gasoline stations, cars, and trucks already designed to use it efficiently; and until about 2005, it was relatively inexpensive.

Without heroic and far-reaching changes, oil is indispensable. Except for a small number of natural gas and electric cars, all our cars and trucks run on oil. Changing our transportation system to use an all-electric or mostly electric infrastructure is possible, but it's an epic (and expensive) undertaking. Coal and natural gas can be converted to oil, but taking this route is clearly a desperation measure—one taken by the Germans late in the Second World War and by South Africa isolated under the apartheid regime. Converting coal to oil is quite expensive and would immediately make climate change *much* worse. Coal and natural gas might give us a bit of time to maneuver, but they won't by any stretch of the imagination allow us to continue the current regime of constant economic growth after oil's decline.[163]

Wood and hydroelectric power are renewable but of little additional help here. Most of the rivers in the world that could be easily dammed are already dammed. Wood can only supply a small fraction of the world's energy needs. In the seventeenth and eighteenth centuries, when the world's population was only about 10 percent of its current levels and standards of living were much lower, the Old World already had difficulty getting enough wood and Europe's forests were rapidly depleting.[164] With a much smaller human population, though—perhaps approaching pre-industrial levels—the situation might be different.

In practical terms, oil is also a stand-in for mineral resources as well as other fossil fuel resources. To access mineral ores, we need energy: to mine the ores, to refine the ores, and to deal with the waste products (the "tailings" of the mining process). Mining and metal processing take nearly 10 percent of the

163 Heinberg, *Blackout* reviews coal.
164 Kwiatkowska, "The Sadness of the woods is bright," p. 42.

world's primary energy, with steel production alone taking about 5 percent.[165] We need energy to get all the metals that make up part of advanced industrial civilization.

The same resource pyramid concept that applies to oil also (and especially) applies to mineral resources. We have logically mined the highest quality ores first; ore quality of numerous popular metals has already been declining for several decades.[166] In the long run, the decline of mineral ores is inevitable and an absolute limit on economic growth. Some common metals (e.g., iron) are sufficiently common in the earth's crust so that this doesn't pose a fatal economic problem. But other metals, once used, are for all economic purposes lost forever. They have dissipated into the broader environment, and it would be prohibitively expensive to recover and recycle them. Because of rising energy costs—rather than because we have exhausted all supplies of mineral ore—we must reckon with finite supplies of a number of metals.

Unconventional oil, economics, and the Hirsch report

Peak oil theorists, including Hubbert, Campbell, and Laherrère, have long been aware of and discussed the possibility of developing unconventional oil. The first significant discussion of the role of unconventional oil in averting peak oil came in a 2005 report commissioned by the Department of Energy, popularly known as "the Hirsch report" after the project leader, Robert Hirsch.[167]

The Hirsch report deliberately made no predictions about the date of peak oil. It merely stated that whenever it *did* occur, peak oil could be tremendously disruptive to the economy, and that to avoid such disruption we needed to prepare at least twenty years in advance. The authors said that "as peaking is

165 Bardi, *Extracted*, p. 114.
166 Bardi, *Extracted*, pp. 114–116.
167 Hirsch et al., "Peaking of World Oil Production."

approached, liquid fuel prices and price volatility will increase dramatically, and, without timely mitigation, the economic, social, and political costs will be unprecedented."[168]

As well as the standard calls for increased fuel efficiency, the Hirsch report advocated development of unconventional oil: heavy oil and tar sands; improved oil recovery, including fracking; and converting coal and natural gas to oil. Developing unconventional oil, for the Hirsch report, was the answer to the problem of peak oil. Whether by deliberate choice or by failure to choose otherwise, during the Obama administration the US economy lurched in the direction indicated by the Hirsch report. Practically all the increases in oil production since 2005 have been in the realm of unconventional oil. This has had numerous consequences such as increased costs; a worsening climate situation; water pollution and earthquakes from fracking; and BP's disastrous *Deepwater Horizon* oil spill. If we gradually switch to unconventional oil as a "solution" to peak oil, then we are moving into a very different energy world: a world of continually rising costs, rapidly increased warming, and accelerated environmental damage.

Conclusions

When will oil peak? Has it already peaked? Because oil is so basic to the economy, that's the question on the minds of everyone who wants to know about peak oil. But because oil is so basic to the economy, it is also likely that *oil will peak only when the economy peaks*. Peak oil won't be a *cause* of limits to growth—the fear on everyone's mind—but rather a *consequence* of limits to growth.

In other words, because oil is so indispensable to the economy—and because the economy is so indispensable to political success—governments will likely intervene to prop up

168 Hirsch et al., "Peaking of World Oil Production," p. 4.

the economy as long as possible. It is only when this becomes impossible, for whatever reasons, that we will see the arrival of peak oil.

We are not going to literally run out of oil anytime soon; rather, the costs (whether political or economic) will just continue to rise. Governments can intervene to keep the oil flowing, but this doesn't truly reduce costs. It just displaces the costs, making them a political or environmental problem rather than an economic one. Increasingly, dirty and more expensive unconventional oil is what we will put in our gas tanks. We need to squarely face these costs and talk about them, not avoid them for fear of panicking investors or the public.

CHAPTER ELEVEN

LIMITS AND THE ECONOMY
The failure of markets

Despite immense environmental destruction and resource depletion, neither our political system nor the economic market seems to be responding adequately. We would expect that scarce resources would rise in price and political leaders would leap into action. But this isn't happening. What's going on?

There are two ways of understanding these failures. The first is through the idea of *market* failures: the failure of our economic system to tell us something is wrong here. The second is through the idea of *political* failures: namely, the failure of our social and political systems to tell us something is wrong and take corrective action. In this chapter, we will deal with the failure of the market; in the next chapter, with the failure of our political system.

Resource depletion and economic failure

If resources our economic system relies on become scarcer, won't this affect our economy? In 1972, *The Limits to Growth* suggested that the earth's resource and population limits could be reached by the middle of the twenty-first century. *The Limits to Growth* proposed multiple scenarios in which, in their view, resource depletion, population, food, industrial production, and pollution would all play out. The "standard model" made projections based on physical, social, or economic relationships. Other projections assumed that world resources are doubled, or stabilizing policies are introduced, or technology is devised to control pollution, and so forth. So far the broad projections in the "standard model" of the original study seem to correspond with our real world pretty well.

But *The Limits to Growth* doesn't address either economic or political failure. It only addresses the end result: the *physical size* of the economy. We could imagine, for example, that the economy doesn't fail at all, but that we see declining human population and declining industrial production through voluntary birth control, the spread of vegetarianism, and a plan of economic degrowth.

Many popular writers in the limits-to-growth community argue that limits to growth *do* imply catastrophic economic failure, and that in the future only small local economies will survive. Conservative social critic James Kunstler comments: "The salient fact about life in the decades ahead is that it will become increasingly and intensely local and smaller in scale."[169] The Post Carbon Institute "promotes the strategy of relocalization—building resilience—through strong local communities."[170] Financial-risk analyst and blogger Gail Tverberg says the economy is a closely knit fabric of interlocking parts, held together by economic growth and huge burdens of private and public debt, which will inevitably fail under pressure of increasing resource scarcity.[171] These writers have important

169 Kunstler, *The Long Emergency*, p. 239.
170 Post Carbon Institute, "Relocalize."
171 Tverberg, "A new theory of energy and the economy – Part 1."

insights, but if we are to look for alternatives, we need to know more precisely *why* markets are failing.

Practices that are unsustainable on national and international scales *prima facie* will also be unsustainable on a local scale. Unsustainably produced goods generally can't be made "sustainable" simply by being produced locally, as this ignores the total energy (not to mention other resources involved) used in production of the product.[172]

Market failures and "open-access regimes"

A market failure is an economic situation where the economy creates the "negative externalities"—unintended bad consequences, like climate change—discussed in chapter 2. Markets may fail for many different reasons. In the textbook *Ecological Economics*, Herman Daly and Joshua Farley devote three chapters just to market failures.[173] If we can judge by the increasingly alarmed statements of scientists about climate change and other environmental issues, it appears that market failures are more the rule than the exception.

Economists call a situation in which *no one* has any property rights an *open-access regime*; the resource in question isn't owned or regulated by anyone. This is an obvious example of a market failure. There's no financial incentive to conserve the resource because no one benefits or could benefit (economically) from its conservation. There are numerous examples of open-access regimes:

1. Whaling in the nineteenth century

2. Open range in the American West in the nineteenth century

....

172 Saunders et al., "Food Miles," p. vii.
173 H. Daly and Farley, *Ecological Economics*, chapters 10, 11, and 12, pp. 157-219.

3. North Atlantic cod fisheries in the late twentieth century

4. Species extinction

5. Gold rushes in the nineteenth century

6. Climate change

In all these cases, the resource in question was or is neither owned nor regulated: whales, grasslands, fisheries, wild species, gold, and the atmosphere. You may as well grab as much as you can because if you don't, someone else will. Garrett Hardin popularized the term "the tragedy of the commons" in an essay by the same title,[174] giving the example of sheep grazing in an unregulated pasture. Hardin's term is a misnomer, though, because the resources in question are not commonly owned. You can deplete or exhaust the resource, but there is no law against taking it because no one owns it.

Other market failures

The problem of natural resources in the economy is deeper than the problem of open-access regimes, though. Even in the case where a physical resource *is* privately owned—thus *not* part of an open-access regime—this doesn't ensure that the resource will be wisely used. Even though markets work quite well for our ordinary purposes (e.g., going to the store and buying food, clothing, or furniture), there are a variety of ways in which they are likely to fail, and these failures are growing more worrisome and problematic with each passing year. For our purposes here, I will mention just three: imperfect information, inappropriate time frame, and future generations.

1. *Imperfect information.* Buyers and sellers may not know key aspects of what they are buying or selling. For example, the price

174 In De Bell, *The Environmental Handbook*, pp. 31–50.

of oil is dependent on many uncertain political factors. No one knows what OPEC is going to do, what some Saudi Arabian oil minister is saying, whether some new Middle East crisis will erupt, or whether a new climate treaty will soon regulate the supply of oil, whether a pandemic will disrupt the economy, or whether a major oil-producing country (like Russia) will precipitate an international crisis. No one, not even the experts, knows for certain how plentiful oil is anyway; many countries are guarding their oil data as a state secret.

2. *Inappropriate time frame.* Even if conserving resources were ultimately profitable, the time frame within which one might realize a profit by conserving resources may be on the order of decades or centuries. For example, farmland suffering soil erosion may not be fully depleted and incapable of supporting crops for another fifty or a hundred years. But how can a farmer, trying to make a living and pay bills just in the next year, profit economically by taking extra steps to avoid soil erosion? The value of that land decades from now is sufficiently far in the future—and the profit sufficiently uncertain even then—that no one bothers.

3. *Future generations.* How will future generations be taken care of? They aren't in the marketplace to begin with. What is the value of depleting groundwater for generations yet unborn? Even if all the previous market failures are resolved, we still have no way to incorporate the value of resources for future generations.

The oil market exemplifies *all* these issues. Increases in the price of oil likely influenced most of the recessions that happened after the Second World War.[175] The economy, in turn, heavily influences the price of oil: if the economy is depressed, the price of oil will be down just because people are less able to

---

175 Hamilton, "Historical Oil Shocks."

afford it. In 2008, at the beginning of the Great Recession, we saw both phenomena: in July oil prices hit record highs, followed six months later by a crash in prices to less than one-third their previous level.

Oil is also a difficult commodity to price for more technical reasons. Demand for oil is *price inelastic*: it doesn't respond quickly to changes in price. People and companies are locked into their habits and don't change quickly. This means it takes a huge increase in prices before demand will be dampened, and a huge decrease to bring demand back. Seemingly minor dips in oil supply can have an outsized effect on prices. Oil companies have found through bitter experience that price volatility is bad for business; if prices vary a lot, people will just stop buying. This is why oil companies try to develop "spare capacity," the ability to rapidly raise (or lower) the output of oil to take care of demand issues. The need for spare capacity is a key reason that oil cartels such as OPEC were developed, so that producers can agree on common action to avoid price volatility.

The end result? The price of oil doesn't reflect its relative scarcity. Instead, it reflects uncertainty and wild variations. If no one can accurately predict the price of oil, then no one has a *market* incentive to leave oil in the ground based on the future scarcity of oil. For the sake of future generations, we should leave most oil in the ground. But that will never happen through the free market—unless oil is left in the ground due to an economic collapse.

Conclusions

When resources are plentiful—as they were when we lived in a relatively "empty world," with very few human structures or domination of the environment—there is little or no *economic* need to worry about resource depletion. From a crassly utilitarian point of view, neither oil, nor greenhouse gas emissions, nor soil erosion, would be much of a concern if there were plenty of oil, plenty of atmosphere to absorb greenhouse gas emissions, and plenty of topsoil.

But when resources are scarce—as they are today—market failures abound. Common-sense considerations imply that the market for numerous natural resources has failed; due to market failures, prices don't reflect the scarcity or abundance of resources. People just can't believe, or accept, the reality of a finite planet. Then, when prices don't reflect a resource-depletion problem, this in turn is used as evidence that we don't have a problem with resource scarcity in the first place! This kind of "head in the sand" mentality multiplies the chances that before we will be able to acknowledge that resource depletion is a problem, we will be staring a full-blown political failure in the face—such as those discussed in our next chapter.

CHAPTER TWELVE

COLLAPSE

The failure of politics

Markets can fail in disastrous ways, but *societies* can fail as well. Our economic system is unlikely to be able to deal with a number of the limits on economic growth. Will our increasingly polarized political system be able to do any better, or will it collapse together with many of the other features of industrial civilization?

The study of how and why societies collapse is unfortunately not a science—at least not yet. Often those who have studied collapse have relied on disciplines such as history or philosophy.[176] Among modern writers who have tried to address the problem of social collapse in an objective fashion, we will look at three: Joseph Tainter, Ugo Bardi, and Peter Turchin. We don't have a timeline of the future, but they can offer us important insights based on the past.

The most famous example of a civilization's collapse is the fall of the Roman Empire, but there are plenty of others, such

---

176 Tainter, *The Collapse of Complex Societies*, chapter 3, gives an excellent overview of the research and writing on this subject prior to 1988.

as ancient Mesopotamia, Easter Island, the Maya Civilization, and Chaco Canyon. Today, some suggest that modern Western civilization could face collapse. Financial-risk analyst and blogger Gail Tverberg argues that we face the collapse of the world financial system, following which much of our industrial system might fail.[177] Population biologists Paul and Anne Ehrlich (and authors of *The Population Bomb*) wrote in 2013 that "Environmental problems have contributed to numerous collapses of civilizations in the past. Now, for the first time, a global collapse appears likely."[178]

The study of collapse

What exactly is *collapse*? Joseph Tainter, a well-known anthropologist and historian, offers us both a definition of collapse and a theory:

> Collapse ... is a *political* process. It may, and often does, have consequences in such areas as economics, art, and literature, but it is fundamentally a matter of the sociopolitical sphere. *A society has collapsed when it displays a rapid, significant loss of an established level of sociopolitical complexity.*[179]

For Tainter, society is a problem-solving mechanism with an internal organization and order. As it develops, it gets better at solving a wide variety of problems and develops a wide variety of specialized ways of coping with them, becoming more complex. As it becomes more detailed and complex, the decision-making process itself, and the resources allocated to that process, begin to consume more and more social resources. At some point, the

....

177 Tverberg, "Deflationary Collapse Ahead?"
178 Ehrlich and Ehrlich, "Can a Collapse of Civilization Be Avoided?", p. 2.
179 Tainter, *The Collapse of Complex Societies*, 4, emphasis in original.

problem-solving mechanism itself consumes so many resources, that it becomes a bigger problem than the problems it was designed to solve. When this happens, society becomes much more collapse-prone and eventually, with increasing complexity, the system will collapse. Tainter argues that resource depletion, in and of itself, doesn't cause collapse:

> One supposition of this view [that resource depletion causes collapse] must be that these societies sit by and watch the encroaching weakness without taking corrective actions.... If a society cannot deal with resource depletion (which all societies are to some degree designed to do) then the truly interesting questions revolve around the society, not the resource. What structural, political, ideological, or economic factors in a society prevented an appropriate response?[180]

If our society were to collapse in response to (say) global warming, from Tainter's viewpoint the interesting fact for any future historians wouldn't be that a changing climate caused our civilization to collapse. The interesting fact would be that *we didn't do anything about it.*

For Ugo Bardi, collapse is a relatively sudden event, which he calls "the Seneca effect." He quotes Seneca as saying "growth is slow, but the road to ruin is rapid." Bardi gives a somewhat different and more general definition of collapse:

> ... collapses ... are always collective phenomena, meaning that they can only occur in those systems that we call "complex," networked systems formed of "nodes" connected to each other by means of "links." A collapse, then, is the rapid rearrangement of a large number of links, including their breakdown and disappearance. So, the things that collapse (everyday objects, towers, planes,

---

180 Tainter, *The Collapse of Complex Societies*, p. 50.

ecosystems, companies, empires, or what have you) are always networks.[181]

Following this idea, we might say that collapse must be a collapse of a specific network, whether a specific social or cultural order, a political order, or a physical structure like a bridge. To say what the impact of "collapse" is on our ordinary lives, or whether limits to growth will precipitate a collapse, we must first be clear on what it is, exactly, that collapses. Using Bardi's language, we might reformulate Tainter's thesis to say that a declining society finds it easier to modify a subnetwork than to redesign the entire network, so systemic problems tend to go unresolved.

Bardi also offers an interesting twist on Tainter's analysis: collapse is not a "bug" or a "problem" of societies at all, it is a *feature*. We should welcome changes that precipitate collapse rather than resist them at all cost; it is our resistance to these changes that induces collapse.

> Indeed, the Seneca effect [the tendency of collapse to be rapid] is most commonly the result of trying to resist change instead of embracing it. The more you resist change, the more change fights back and, eventually, it overcomes your resistance. Often, it does this suddenly.[182]

Bardi later develops his theoretical model by explaining that resource depletion itself doesn't cause collapse. Collapse wouldn't occur because of resource depletion per se, but only if we try to *resist* the physical reality of resource depletion, rather than becoming more resilient and accommodating these overall limitations while preserving the basic structures of society. He doesn't offer any quantitative or specific recommendations to predict or to avoid collapse; he offers general arguments that

---

181 Bardi, *The Seneca Effect*, pp. 1–2.
182 Bardi, *The Seneca Effect*, 3.

it may be possible to describe social structures that were more resilient and therefore more able to deal with collapse.[183]

Cliodynamics

Tainter and Bardi don't attempt to quantify their theories. In fact, Tainter questions whether this is even possible. Ancient societies didn't keep detailed enough records to make the case either way;[184] "the framework for explaining collapse could probably not be subjected to a formal, quantitative test".[185]

Peter Turchin and others in the newly developed school of *cliodynamics* seem bent on proving Tainter and Bardi wrong on this point and providing empirical historical data in support of their theories. Cliodynamics is the study of history using quantifiable empirical data to address historical questions. Turchin's book *Secular Cycles* (co-authored with Sergei Nefedov) addresses the problem of collapse by looking at the entire cycle of a civilization with empirical data.[186] His later book, *Ages of Discord*, looks specifically at the United States.[187] The general conclusions are that social decline or collapse is fueled by a convergence of reinforcing structural and demographic factors: increasing population, popular immiseration, elite overproduction, and state instability. Turchin acknowledges his debt to Marx (with respect to the structural factors of class conflict) as well as Malthus (with respect to the demographic factors of population increase and declining wages).

....

183 Bardi, *The Seneca Effect*, chapter 4. p. 145: "Is there an 'equation of resilience' that could be applied to a generic network? Apparently, not. Nevertheless, progress is being made in this direction."
184 Tainter, *The Collapse of Complex Societies*, p. 127.
185 Tainter, *The Collapse of Complex Societies*, p. 192.
186 Turchin and Nefedov, *Secular Cycles*.
187 Turchin, *Ages of Discord*.

Roughly interpreted, as a society prospers, the population expands and the number of laborers increases. But because of the increase in population and therefore the increase of laborers, wages fall. Thus, the condition of the ordinary people (workers, peasants, or slaves) necessarily worsens ("popular immiseration").

Worsening economic conditions aren't an intrinsically unsolvable problem. But, Turchin and Nefedov argue, the ruling elites do nothing to address it. It's not that they *can't* react, but that they *don't*. The elite class (landlords, capitalists, or whatever elite exists in the society) *benefits* from the low cost of labor: they can hire labor at a cheaper price and increase their profits. From their point of view, things are going great. What is there to correct?

Naturally, this doesn't end well for the elite class in the long run. More and more people, seeing the profits the elite class makes, want to join them ("elite overproduction"). Soon it is the elite class that is bloated, and the elites are in competition with each other. At the same time, tax revenues fall because of falling wages and a diminishing tax base. At some point, conflict will decrease the numbers of the elites, possibly through a civil war or revolution. A new elite class will emerge on top—either on its own or possibly in alliance with the lower classes—and a new cycle will begin.

Secular Cycles examines eight secular cycles: those of both Republican Rome and Imperial Rome, and those of medieval and early modern France, England, and Russia. Turchin introduces a variety of statistics in support of his theory, showing that toward the end of a cycle, the average price of food tends to increase relative to the average wage of a laborer; the number of the elites in society (whether senators, knights, or nobles) tends to increase; and social instability and conflict also increases. In *Ages of Discord*, he makes adjustments to his model for the modern world, reflecting technical and scientific advances that influence the demand for labor, but leaving the overall theory roughly the same.[188] In this way, cliodynamics has brought a quantitative element to historical analysis of societal decline or collapse.

Will our society collapse?

Do these ideas give us any insight into whether our society will collapse? Briefly, yes.

Turchin's ideas are the most immediately relevant to our subject. His structural-demographic theory has a firm empirical basis, and Turchin's 2010 letter in *Nature* about increasing political instability in the United States now seems marvelously prescient.[189] The threats to our society do indeed seem to come from the directions Turchin is indicating: there are too many elites, relative wages are stagnant or declining, the state is weakening with fewer and fewer revenues coming in from cash-strapped taxpayers, and corruption is rampant throughout society and especially politics.

Anyone who has followed political news over the past few decades knows we have serious internal discord in the United States right now. Many of the features of declining societies to which Turchin draws attention are present in our own society in objectively measurable ways. Total economic activity (reflected in GDP) has increased steadily in the past fifty years, but the benefits have gone overwhelmingly to the rich. While real wages have *declined* slightly over the past fifty years, the number of multimillionaires ("elites") has increased rapidly.[190]

188 He assumes a "monotonic trend to high popular well-being resulting from scientific and technical progress" (*Ages of Discord*, p. 14) but then oscillations in the demand for labor based on other factors, including "cultural" factors (chapters 9 and 12).

189 Turchin, "Political instability may be a contributor in the coming decade."

190 Turchin, "A History of the Near Future," slides 6 and 14.

In the meantime, the inflation-adjusted federal minimum wage peaked in 1968 and has declined since then.[191] We have increasing political polarization, declining levels of social trust, and a weakening state sinking further and further into debt.

But is any of this connected to resource or environmental issues? Turchin never really confronts this question. Intuitively, it seems obvious that declining resources would mean social instability, at least at some point, but Turchin doesn't assign any particular importance to resource-depletion issues, except to give them a minor role.[192] In *Secular Cycles*, expanding population is the driver behind social disorder. There are too many workers chasing too few jobs, thus "labor oversupply," and thus falling wages.

But there *is* a connection between resource depletion and social instability in Turchin's structural-demographic ideas. The connection lies in the impact of population on labor oversupply. Increasing population doesn't always create "labor oversupply"; obviously in America in the twentieth century (up until about 1970 or so), we saw overall increasing population *and* rising living standards. In fact, Turchin himself supplies a counterexample and an explanation. In England in the mid-seventeenth century, population increased *and* wages increased at about the same time, because of dramatic gains in agricultural productivity. When land productivity *increases*, food resources increase, and population pressure on resources therefore decreases, even in the face of increasing absolute population numbers. The increase in population is more than offset by the

....

191 Wenger, "Working for $7.25 an Hour."
192 Turchin says, of environmental impacts on Mayan civilization: "Clearly the resource base of the society is important and climate fluctuations affect it. But it's a secondary factor – a modifier, rather than a prime mover." Turchin, "Collapse of Complex Societies: Did Drought Kill off the Mayans?" Note: Turchin is probably right about the Maya.

increase in productivity.[193]

Turchin suggests, therefore, that the real driver is not absolute population numbers, but *population pressure on resources*. This increasing *pressure* on resources is strongly associated with declining wages, which in turn leads to all the destructive social phenomena that we now see around us: the proliferation of elites, political instability, and so forth.

In pre-industrial societies, such as those discussed in *Secular Cycles*, the primary resources are pretty much fixed and technology is always about the same; "resources" just means "land." Increases in population really *did* mean Malthusian-style popular immiseration. But in the modern highly industrialized United States, limits on land have almost completely disappeared as a factor in popular immiseration. We now have so much agricultural output that most Americans are eating meat like medieval kings and queens. Eventually, driven by such things as soil erosion, land *will* become a problem again, but not just yet. In the meantime, we're now busily exporting our destructive lifestyle to China, India, and Japan.

What *is* an immediate limit on the economy is our consumption of fossil fuels, especially oil. As we saw in chapter 10, episodic oil disruptions were associated with almost all our recessions since the Second World War. Since about 2005, oil has become a chronic problem for the economy, with historically high prices. The situation today is exactly the reverse of that in sixteenth-century England, when increases in agricultural

....

193 Turchin, *Secular Cycles*, pp. 108–110. Starting in about the mid-1600s in England, population increased but real wages *also* increased in correlation with an increase in agricultural productivity. Turchin notes: "population pressure and inverse real wages fluctuated virtually in perfect synchrony," p. 110. This discussion of "population pressure or resources" reappears in *Ages of Discord*, p. 13, but thereafter "resources" seem to disappear as a topic.

productivity decreased demographic pressure and allowed wages and population to rise together. "Population pressure on resources" can be altered either by variations in population or by variations in resources. We need to get a handle on both parts of this equation.

We suspect that our current social crisis is at least partially due to resource shortages and ecological destruction, which are now beginning to intrude on day-to-day social and economic realities. In an alternative universe in which there was plenty of oil, and burning this oil miraculously did not cause global warming, the standard of living would be higher and political instability would probably be less severe. There might never have been an invasion of Iraq, the Great Recession would have been less severe or might not have happened at all, and so forth.

Other factors such as disease (chapter 17), population (chapter 18), as well as automation and wealth concentration (chapter 21), have also played a role in popular immiseration. But the real economic consequences of environmental destruction, which chip away at the very basis of the economy and all life on Earth, still lie mostly in the future. As these consequences begin to bite, we may expect the processes fueling the disintegration of our society to accelerate.

How could we respond to collapse?

Do these ideas give us any insight into how we should respond to a potential collapse? The one major complication here is that we don't know exactly what "collapse" means. What collapses? Bardi suggests that what collapses is always a network, but which network?

We might see an economic collapse like the Great Depression of the 1930s. Alternatively, collapse might be political, and the United States might break up into smaller political units (perhaps along the lines of "red states" and "blue states") or disintegrate into civil war. But we might also see a technological collapse and a reversion to nineteenth-century technology. The US government, whose foundations are based in the eighteenth century, might continue to exist even in what would be, for current generations, a more technologically or economically primitive setting.

To respond to the environmental crisis will require not only awareness of environmental problems and presence of mind, but also a much greater degree of social cohesion than we currently have. That cohesion starts with dramatically reducing the power of the elites. We need to restore some sense of cooperation and mutual trust to our society, and that cannot take place if there is blatant and growing social injustice.

All three of these thinkers sound one optimistic note. They all agree that resource depletion *in and of itself* did not cause the sort of internal failures that we associate with Rome and the Maya. For that to happen, some kind of *societal* failure had to take place. Nothing was predetermined by fate or by resource shortages. Turchin comments that while the onset of the crisis is broadly predictable (weakening state, too many corrupt elites, growing inequality), the outcome is highly *unpredictable*.[194] The future may look bleak, but the outcome really is in our hands.

Conclusions

In Isaac Asimov's classic science-fiction novel *Foundation*, the author describes a future galactic civilization. In the novel, a key character—Hari Seldon—is able through scientific analysis to foresee the downfall of galactic civilization and the need to build a structure that would preserve the civilization's knowledge. Seldon turns out to be right about the collapse, but (not to spoil the plot too much) even he fails to see some key events.

---

194 Turchin, "The Ginkgo Model of Societal Crisis."

We still can't predict the future, but we do have some quantitative data on the sorts of problems that engender collapse. But, as with Hari Seldon, there are doubtless things we can't foresee. Most critically, we don't know what our own reaction to collapse might be and how it might transform the future.

We should be careful what we wish for. There are many paths to an unpleasant (for humans) conclusion to the environmental crisis. There may also be several paths to a solution that involves embracing limits. But right now, we have yet to identify even one.

PART 3
MOVING TOWARD AN
ECOLOGICAL CIVILIZATION

CHAPTER THIRTEEN

PARAMETERS FOR AN ECOLOGICAL CIVILIZATION

Implementing degrowth

Where do we want to go? We want to get to an ecological civilization, but what exactly is that, and what would it look like?

Answering this question is often derailed by thoughts on a second, related question: "How do we get there?" Degrowth, let alone "limits to growth," is a marginal issue on the current political scene. It's easy to get depressed about these environmental issues—and the lack of awareness around them—and never get to the point of thinking about what an ecological civilization would look like. This has led to a minor proliferation of "doomers"—people who believe that civilization is inevitably going to collapse. They will follow our analysis up to this point and conclude: "And then, civilization collapses. The end."

Given all the environmental destruction our economy has precipitated, perhaps what we should really fear is our economy *not* collapsing. Before we get distracted with fears of impending destruction, let's look at some basic parameters for where, exactly, we *want* to go: an ecological civilization.

Identifying parameters

What would an ecological civilization—a sustainable human civilization in balance with the ecosystem in which it has evolved and on which it depends—look like? What are the limits we should suggest that humans should embrace?

Identifying *specific* parameters for a sustainable civilization is harder than it might seem. There are many questions that we can't fully answer in the space of this book—questions, in fact, that even the experts don't have answers for, or can't agree on. But just because we can't answer every question with the precision we'd like, doesn't mean we can't say anything. If we wish to move toward an ecological civilization, the evidence all points our current economy in the same direction: smaller, often a lot smaller.

The coming chapters will be spent examining these parameters for a sustainable civilization. As we do so, we will also start to consider the related question, "How do we get there?" that we are momentarily postponing. In this chapter, I want to give a quick overview of what these parameters are and why I think the economy is going to be much smaller than it is today.

The parameters I am considering fall into two broad (and sometimes overlapping) categories: energy issues and biological issues. Energy issues encompass energy supply, renewable energy, nuclear power, and the availability of minerals and metals. I subsume minerals and metals under the general category of energy, because today energy is our key limit on acquiring minerals and metals. Biological issues include biodiversity, world hunger, food, agriculture, and the all-important question of human population.

1. *Energy issues.* We need to sharply limit or eliminate fossil fuel use, of course, both for climate change and peak oil

considerations. Climate considerations indicate that we should limit fossil fuel use ourselves; peak oil considerations suggest that Mother Nature will soon do it for us. It's an open question which limit will decisively bite the economy first. From what we've seen of political intransigence and scientific illiteracy, my bet would be on peak oil, which likely arrived several years ago, and which has already caused (or at least contributed to) the Great Recession. But we also seem to have blown past a lot of climate tipping points already, potentially setting us up for the worst of both peak oil disasters and climate disasters.

In theory, limiting fossil fuel use doesn't mean limiting total energy use. We've got renewables, hydroelectric power, and nuclear power to fall back on. But in practice, it does mean lower energy supplies. There are no quick, easy, and plentiful substitutes for fossil fuels. It's certainly possible that by 2050 we will all be sipping lemonade underneath solar panels and wondering why we all made such a fuss about the viability of renewable energy early in the twenty-first century. But it's more likely that we will still be struggling with serious energy-supply problems after accounting for variables like metals and minerals, industrial heat, energy storage, electrification of the economy, and the electric grid (the infrastructure, such as transmission lines, needed to transmit power from its origin to the consumer). Renewable energy technology is much more problematic than most of us imagine, and unfortunately we haven't yet had any robust public discussion of issues concerning the viability of renewables—all of which will be further discussed in chapter 14.

The main alternative to renewables—that still avoids carbon dioxide emissions—is nuclear power. If we are truly committed to avoiding carbon dioxide emissions, and renewables stumble in providing the energy we want, nuclear will be the key alternative. Nuclear is rather unpopular with the public but so far has proved quite a bit safer than most of us think. Nuclear might be able to help sustain a much smaller economy for a long time, perhaps thousands of years. Even so, we don't know how much nuclear fuel we've got, nor do we know how well nuclear can provide industrial heat (the very high temperatures required by heavy industry). Nuclear advocates typically don't think in terms of embracing limits; they think in terms of supporting an ever-growing economy, so the question of how much nuclear fuel we have has not been thoroughly researched.

Interestingly, nuclear power and renewables such as solar and wind power share one common problem: an electrification problem. They all generate electricity, yet many ordinary devices that require energy aren't electric: cars, except for a small number of electric cars, being an obvious example. However we want to handle this, it will take energy and resources to get our devices to use that energy.

Biofuels, including wood, biodiesel, and biogas, are the one truly renewable-energy source that could answer a lot of the objections to nuclear and modern renewables (such as the ability to provide industrial heat). However, biofuels require *so* much agricultural land that they couldn't support more than a fraction of today's economy. Biofuels, however, might be able to support a much smaller economy, and so we would want to know: exactly how large an economy could biofuels support?[195]

Our overall supply of minerals and metals are finite, just like fossil fuels. So far, the exact extent of minerals and metals in the earth's crust hasn't mattered; the limiting factor has always been the energy required to blast them out of ores of decreasing quality. But if we have a spectacular energy breakthrough (say, nuclear fusion?), suddenly the absolute supply of metals and minerals might become a limiting factor after all. Some common metals may be in sufficient supply to last for thousands of years, tens of thousands of years, or even longer—but only in a much smaller economy.

195 Cheerfully discussed in Alice Friedemann's book *Life After Fossil Fuels.*

So, realistically, how much energy could a truly sustainable economy provide? All these caveats leave us with an extraordinarily broad range of sustainable energy futures. It could be "four billion humans with something like a European lifestyle today" to "a much smaller human population of 100 million with an economy running on biofuels"—or something in between, or something else altogether. In coming chapters, we'll try to show why precision on this question is so difficult to find, as well as try to narrow down the range of possibilities somewhat.

2. *Biological issues.* In the long run, it's *biological* issues such as biodiversity, agriculture, and population, that are the fundamental limit on the economy. Without oil, we can always go back to wood and the horse and buggy. But without sufficient soil, or in the face of biodiversity collapse, we are looking at human extinction or reversion to a *very* primitive level of human existence. Hunting and gathering wasn't working all that well ten thousand years ago; even at its height, perhaps thirty to forty thousand years ago, it couldn't provide even a fraction of today's standard of living.

Soil erosion is a critical problem, as soil is eroding ten to twenty times faster (at best!) than natural soil formation. Here we have a specific metric: back-of-envelope calculations suggest that even if we were all vegetarians or vegans, the maximum human population that our food supply could support is on the order of one to two billion. That would imply keeping about 90 to 95 percent of today's cropland fallow, allowing soil to recover while we farm a small percentage. Similarly, water problems suggest that ultimately we won't be able to maintain the yields of the Green Revolution and that much of today's agricultural lands (e.g., the US Great Plains) will be out of production in about a century. Optimistically, though, were we able to all farm organically and implement permaculture or no-till techniques, we might considerably reduce soil erosion—though we don't know by how much or with what yields.

Even if we solve all other problems, biodiversity collapse is a deeply troubling "wild card." Conceivably, we may have *already* "pulled the trigger" on human extinction because extinctions can cascade in unexpected ways. The decline in insects will likely trigger a decline in insect-eating birds, and so forth, and these declines may eventually work their way up to *us*. The same problems that emerged in Biosphere 2 (a decline in insects and other pollinators) are now unfolding in *our* biosphere. We need to mitigate the causes of extinctions, especially human land use. On the plus side, meeting other limitations (e.g., on agriculture and energy) may greatly lessen human impact on the rest of the biosphere in any case. If there are fewer humans cultivating less land, there will likely be more reclaimed wilderness.

Climate change is ultimately a biological problem as well as an energy problem because about half of the climate problem is due to our destruction of the biosphere and our heedless, vast multiplication of livestock.

What the future looks like

So is civilization doomed? Should we bid a fond farewell to civilization and wish everyone good luck in the hardscrabble, violent subsistence world of hunger, disease, and war to follow?

Despair over the climate and the environment is spreading. We certainly shouldn't give in to despair, but what exactly are we offering instead? Unless there is a factual recognition of the utter seriousness of the situation and a response proportional to the problem, we haven't really addressed either the environmental crisis itself or the problem of environmental despair.

The shrinking of our economy is inevitable. It *will* get smaller. For most of the biosphere, which has been devastated by human economic activity for at least the past century, this is excellent news. Whether the shrinking of the economy means a "collapse" of some human social, political, or economic order is less important than whether this can happen as compassionately as possible—which is exactly what the rest of this book will discuss. Within the framework of embracing limits, there is nothing inevitable about our fate or the future of human civilization.

Once we look systematically at each of the problems we've identified and begin to address each one in radical fashion, we will see that their solutions tend to converge. They converge in the common-sense approach I outlined in the introduction: (1) substantially reduce personal consumption, (2) substantially reduce human population, and (3) eliminate or drastically reduce livestock agriculture. We need to do all three things: half measures are not enough.

We don't know exactly what a sustainable future will look like. There are still numerous imponderables and complexities here: questions of fact that even informed, objective observers can't yet answer. But there are some things of which we can be relatively certain. In the Middle Ages, and indeed well into the eighteenth century, civilization was fully sustainable without fossil fuels. Most of us wouldn't feel comfortable with the standard of living of the Middle Ages, or of the eighteenth century, even as kings or queens, but this level of civilization at least represents a kind of "technological floor" for what is sustainable. Furthermore, with modern science we've figured out many technologies that don't require enormous quantities of energy: medical advances such as anesthesia, antibiotics, and birth control; simple technology like washing machines, plows, and tractors; basic educational facilities and books; and more. Surely, we could add these advances to our technological floor without unduly straining the earth's resources, at least at some level of human population. It may be helpful, conceptually, to think of how we can sustainably *add* to this technological floor, which we can be reasonably certain is sustainable, rather than what we must "give up" (phones? cars?) to get to sustainability.

In the science-fiction time-travel romance series *Outlander* (minor spoilers here), our heroine (Claire) is accidentally thrust back into the eighteenth century. Claire seems to manage quite well in the eighteenth century because she was trained as a nurse in the Second World War; the one medical advance that she needs the most is antibiotics. In one of her trips, she manages to bring small amounts of antibiotics with her, and it also proves quite useful that she was vaccinated against smallpox in the twentieth century.

We might ask ourselves: What would we really need to make the world of the Middle Ages, or the eighteenth century, tolerable and worthwhile? If we can imagine that, then we can tackle the problem of "reverse engineering" civilization to get to that point. Reaching a social consensus on such issues seems impossible today, but as the cascading crises of the twenty-first century become increasingly obvious, what is impossible today may be possible tomorrow.

We must reduce current levels of world population. Agriculture and inequality are the key limitations on population. Not everyone has to consume at today's inflated American levels, but everyone needs to eat, and this establishes a floor beneath which we can't allow food consumption to sink. Soil and water issues make long-term sustainability of our current food system impossible. Soil erosion at ten, twenty, thirty, or forty times the rate of soil formation isn't sustainable even if everyone were vegan. Water is indispensable also, but many of the most agriculturally productive areas of the world (such as central California!) rely on depleting supplies of groundwater. Even without these agricultural issues, raising eight billion people even to the standard of living of European countries like France and Sweden, much less to the bloated American standard of living, would be a colossal and environmentally destructive task; sustaining it over the next few hundred years is almost certainly impossible. Human population probably needs to decline, but it isn't an immediate problem; by addressing the overpopulation of *livestock* we can take care of much of the "population" problem.

We must drastically reduce or eliminate the livestock industry. We must come down hard on livestock agriculture. From a technical point of view (leaving behind social and cultural obstacles for the moment), livestock animals don't really benefit us. Eating them is harmful to human health, and raising them is catastrophically destructive of the environment. Though beef lovers will react with horror, when we consider only health and environmental issues, drastically reducing livestock agriculture lacks a significant downside.

The single major exception in eliminating livestock *may* be for animals used for farm labor and transportation, if it proves impossible (due to resource limits) to support any sort of advanced industry such as building tractors and railroads. How we should treat animals (raised for labor or for food) raises a host of other social and ethical questions that we can't resolve at this point in the book. For now, all we will say is that our society's view of "nature" (plants, animals, insects, fungi, minerals, or whatever) needs to be addressed somehow.

Drastically reducing the livestock industry is *technically* the easiest of these three guideposts to follow, even though for practical reasons it is a difficult step for many people to take, even the environmentally conscious. Universal or near-universal veganism would hardly compromise our standard of living: it requires minimal new expensive machinery (unlike building out a renewable-energy infrastructure) and can be implemented quickly (unlike nonviolently reducing human population). It could help curb the biodiversity crisis by allowing reforestation and providing additional habitat, and it would improve human health.

We still have a long way to go; there is much else related to the biodiversity crisis besides livestock agriculture. But getting rid of livestock agriculture is the "low hanging fruit" of our efforts. Because livestock are so destructive, it is quite unlikely that we can address other problems with the biosphere without a vegan or near-vegan economy, in which plant-based diets are mostly or entirely accepted as a cultural and ethical norm.

We must reduce levels of personal consumption. "We" here refers to those of us in the industrially advanced countries of the world; less developed countries likely need to increase their levels of consumption. We distinguish the industrial sector of the economy, that which produces non-food consumer goods, from the food sector. We can't maintain current US levels of personal consumption over the long run, and as we will see, it will be even more difficult to "raise" the rest of the world to this destructive standard of living. The metals, minerals, and fossil fuels to do this over the long haul simply don't exist. We need to reduce total consumption, including by eliminating most or all fossil fuels. Despite brave talk about a "Green New Deal," building a renewable infrastructure will be an epic undertaking that will require what many will regard as "sacrifices" in our standard of living and is unlikely to ever create the energy output of our fossil-based industrial system. That doesn't mean that we can't have a worthwhile human civilization and indeed be even happier than we are today. But we can't judge how worthwhile or meaningful human life could be based on rigid and arbitrary comparisons to a level of energy output or production of material goods.

Integral to all these efforts is another key theme of this book: the path to this sustainable future goes through the doorway of social justice—and even deeper than this, through the doorway of ethical understanding. The drastic measures I've outlined above will doubtless eliminate many people's jobs or other sources of income. The burden of implementing these measures must fall on rich countries and on the rich individuals in each country. Everyone should be adequately supported in a simpler lifestyle, which is where most of us will be. There will probably be an "upper class," but there will be a maximum income and this "elite class" won't remotely resemble today's wealthy.

Conclusions

While uncertainty about the future makes precision impossible, we still know what essential changes we need to make and roughly where the uncertainties exist. In the coming chapters, let's look at these parameters and see how they might work. Here's what comes next:

1. New sources of energy to replace carbon-based fossil fuels (chapters 14 and 15).

2. New biological relationships with nature, such as reduced human population, plant-based diets, and devoting half the planet to wilderness (chapters 16, 17, and 18).

3. New social policies to limit the scale of the economy and change the distribution of wealth (chapters 19, 20, and 21).

4. New cultural systems to accommodate our changed relationships with ourselves and with nature (chapters 22 and 23).

5. Making the transition to an ecological civilization (chapter 24).

We can't see our path forward precisely. But we know enough already to say that *half measures are insufficient.*

CHAPTER FOURTEEN

RENEWABLES TO THE RESCUE?
The limits of renewable energy

Renewable energy addresses climate change, peak oil, and fossil fuel depletion generally. Renewable energy typically refers to solar photovoltaic (solar PV), concentrated solar power (CSP), and wind power. But other renewable sources also exist, such as geothermal energy, tidal energy, hydroelectric power, and biomass energy (typically wood). Wood was the dominant energy source for the world until a few hundred years ago, and it remains so in some parts of the world even today.

Modern proponents of renewables sometimes state that not only can we make a transition to a renewable-energy economy, but we can also still have a *growing* economy.[196] Mark Jacobson and Mark Delucchi have said that low-cost, reliable systems that replace 100 percent of fossil fuels with renewable energy could solve our energy and climate problems relatively easily.[197] In

196 International Energy Agency, *Perspectives on the Energy Transition*, p. 12: "The energy transition can fuel economic growth and create new employment opportunities."
197 Jacobson and Delucchi, "A Plan to Power 100 Percent of the Planet with Renewables."

February 2019, Representative Alexandria Ocasio-Cortez and Senator Ed Markey, US Congress, introduced resolutions for a "Green New Deal" to develop renewable infrastructure through a ten-year national mobilization.[198]

Renewables are important—no doubt about it—but renewable energy does *not* negate limits to growth. Even if we have plenty of renewable energy, there are still limits on biomass, population, soil, water, minerals, and metals.

But there is a deeper and more troubling problem here. Renewables aren't likely to work nearly as well as modern proponents imagine they will, and there is little evidence of any "national conversation" about what we can do about that. Distinguished energy scientist Vaclav Smil says that because of the scale involved, the goals we've set for ourselves to reduce carbon emissions are "delusional."[199] This doesn't mean that a renewable economy is impossible or that renewables aren't worth pursuing, especially if the alternative is climate catastrophe. It just means that we should be honest with everyone about what renewables can do and what they will cost.

Evaluating energy sources through EROEI (energy return on energy invested)

EROEI, or "energy return on energy invested," is a way of measuring the usefulness of an energy source. It takes energy to produce energy. To get oil, we need energy to drill the oil well, refine the oil, and get it to gas stations around the world. Fortunately, the energy we get from burning oil is many times greater than the energy it takes to drill oil wells and get it to gas stations around the world. In the early days of oil drilling,

....

198 Kurtzleben, "Rep. Alexandria Ocasio-Cortez Releases Green New Deal Outline."

199 Marchese, "This Eminent Scientist Says Climate Activists Need to Get Real."

the EROEI of Texas oil wells could have been as high as 100 to 1 (that is, the energy returned was 100 times greater than the energy invested). The EROEI of modern corn ethanol, by contrast, is much lower—on the order of 1.4 to 1 or 1.2 to 1, and even this depends on some dodgy accounting of the "energy return."

EROEI research is still new. While most agree that EROEI and the concept of energy surplus is important, we don't have clarity about how to measure EROEI, nor do we know what high or low EROEI values imply for society. In medieval Europe, EROEI was low. The "energy" was mostly food energy, animal labor, and wood for heating. Most people were farmers who only generated a small surplus of food and energy beyond that required to feed themselves. This small energy surplus provided support for a small number of kings, nobles, monks, and nuns.

Today we have a tremendous energy surplus and only a small minority of people are farmers. Our modern surplus allows us to provide such things as universal public education, health care, roads, a court system, and police. If the EROEI of our society falls too low, our energy surplus will be less, and we may not be able to afford these things. Perhaps we will only be able to provide a sixth-grade education and a limited transportation infrastructure for most people, for example.

So what is the EROEI of renewables compared to, say, the EROEI of oil? Mason Inman surveyed a number of studies and came up with a value of 20 for wind, 6 for solar PV, and 40 for hydroelectric. Hydroelectric return is excellent, but we've already utilized most of the rivers that could be dammed. The EROEI of wind is comparable to modern oil production (which Inman estimates as 16:1). However, Inman's estimates did *not* incorporate the energy lost in building energy storage for renewables.[200]

200 Inman, "How to Measure the True Cost of Fossil Fuels," p. 59: "The [EROEI] measure does not evaluate all the benefits

Neither the wind turbines nor the energy storage are permanent; they will eventually break down and need repairs or replacement. All this electric energy will need to be delivered to the end users (consumers) via the "electric grid," which includes all the transmission lines, electric generators, and distribution networks required to enable end users to turn on the light switch and expect the light bulb to light up. But because renewables generate electricity, this electric grid will need to be greatly enlarged. The need for energy storage; the energy and materials costs of repairing, expanding, and then maintaining our antiquated electric grid; and various other technical needs, all tend to chip away at the EROEI of renewables. And if the EROEI of renewables falls to less than 1:1, that means a net energy return of less than zero, which is worse than useless.

1. Intermittency and storage

What do you do when the sun doesn't shine and the wind doesn't blow? Solar power obviously doesn't work well at night unless it's been captured and stored somewhere during the day. The intermittency of wind power is a serious problem. Periods of almost complete calm aren't rare. One study modeled energy availability from wind in Great Britain and found that during some periods of maximum electricity demand (in the winter), there would be almost no wind power generated. Any storage solution proposed for Great Britain would need to store several days' worth of energy for the entire country.[201]

There are different ways of addressing this issue: overbuilding renewables, transmission lines linking different parts of the world, and energy storage. Pumped hydroelectric storage with

---

and drawbacks of a fuel; notably, it does not address the ... intermittence of wind or of solar power."

201 Oswald et al., "Will British Weather Provide Reliable Electricity?"

surplus wind or solar energy (pumping water upstream of a dam and letting it flow through the dam later to generate electricity) is reasonably efficient, but there aren't many sites where this sort of pumping is possible. Using batteries to store renewable energy for cloudy and calm days is technically feasible, but the undertaking is gargantuan. Backing up *one day* of US electricity generation using sodium sulfur (NaS) utility-scale batteries would cost more than 40 trillion dollars, cover 923 square miles, and weigh 450 million tons. However, advanced lead-acid batteries would do quite a bit better, costing about 9 trillion dollars, covering about 217 square miles, and weighing just under 16 million tons. Then, when the batteries give out in a decade or so, we can start over again from scratch.[202]

Batteries take up much more space to provide the same energy currently provided by fossil fuels. Today, the diesel fuel that powers large long-distance trucking is on the order of 60 to 120 times denser than lithium-ion batteries.[203] We could hope for technological improvements, but during the past century, the energy density of batteries has only increased about sixfold since the first lead-nickel batteries.[204] It will be quite difficult to design electric trucks for long-distance hauling; the weight of the batteries is an important obstacle.

Another possible system is "power-to-gas": using surplus renewable electricity to generate hydrogen, methane (CH_4), or some other gas, which can then be stored and used at our leisure. It would be possible, for example, to use wind power to create methane from CO_2, which could be used as an alternative to conventionally extracted natural gas and delivered in already existing natural gas pipelines.

Methane and hydrogen both have their own problems. Right now, the whole energy-generation process of using renewables

202 Friedemann, *Life After Fossil Fuels*, p. 77.
203 Friedemann, *Life After Fossil Fuels*, p. 42.
204 Friedemann, *Life After Fossil Fuels*, p. 41.

to generate methane has an efficiency of about 34 percent,[205] which isn't great, although with new technology this number might improve. Methane is, of course, a greenhouse gas, and it requires inputs of CO_2, so its use doesn't free us from fossil fuels or climate change. However, it could alleviate supply problems due to peak oil and could work using existing natural gas pipelines. Hydrogen is cheaper to produce than methane and isn't a greenhouse gas. But it's much more difficult to store and burn and would likely require completely new infrastructure to store and transport it,[206] all of which would chip away at EROEI.

No matter what we do, intermittency means that we will get less energy for each solar panel or wind turbine built than the absolute amount of energy generated at the source. Intermittency isn't a *fatal* problem for renewables, and many people are working on new technologies to address this problem. But any storage technology will diminish the EROEI of renewables, increasing both energy and financial costs to create the final energy output.

2. An all-electric economy: infrastructure issues

Solar and wind power create electricity. (In fact, so does nuclear energy.) The Hirsch report, described in chapter 10 on peak oil, already highlighted a key problem caused by oil shortages. The immediate problem with peak oil isn't a shortage of *energy*, but a shortage of *liquid fuels* that our cars, trucks, and factories can process—namely, oil. An all-renewable economy would most likely be a nearly all-electric economy as well. To begin with, everyone will have to switch over to electric cars and everything that this switch implies, requiring an epic ramping up of battery technology or hydrogen- or methane-based cars.

205 Utrecht University Faculty of Science, "Methane promising route for storage of renewable energy from sun and wind."
206 Utrecht University Faculty of Science, "Methane promising route for storage of renewable energy from sun and wind."

It is technically possible to do this, but it is neither quick nor inexpensive. Electric cars currently have a restricted distance range, and the most rapid battery recharge currently takes several hours. There are proposals for gas stations to become "electric battery stations" instead; they would swap car batteries in and out of cars, replacing an exhausted car battery with a fully charged battery. Electric cars today often use lithium-ion batteries, but large-scale use of lithium-ion batteries in cars could create lithium shortages.[207] Large vehicles such as heavy-duty trucks, which can weigh forty times more than an average car, probably couldn't work on a stand-alone electric basis at all.[208] For long-distance hauling, we would need to shift to electric trains or electric trucks powered from overhead transmission lines.

There's another problem with electricity: it's difficult for electricity to support heavy industry. Heavy industry often requires "industrial heat"—temperatures that are very high, steady, and continuous. For example, conventional steel blast furnaces operate at 1100 degrees Celsius. Heavy industry produces concrete, steel, and petrochemicals and is responsible for about 22 percent of global CO_2 emissions.[209] In an all-electric economy, creating these high temperatures would be quite expensive. The relatively best option from a *climate* perspective might be continuing to use fossil fuels but using carbon capture and storage for the carbon dioxide emissions. Carbon capture and storage would address the climate problem but would still rely on finite fossil fuels and so wouldn't be 100 percent renewable. Another currently more expensive option is using hydrogen as fuel. Hydrogen produced from renewable electricity

207 Vikström et al., "Lithium availability and future production outlooks."
208 Friedemann, Alice, "Diesel is finite."
209 Friedmann, S. Julio et al., "Low-carbon Heat Solutions for Heavy Industry," p. 7.

is the "cleanest" solution but would, with today's technology, increase costs by 200 to 800 percent.[210]

This raises the difficult question: Will it ever be possible to produce renewable energy from only renewable sources? Might we ever see the day when a wind turbine can be manufactured with all the energy in its manufacture, including mining the metals needed, coming from renewable energy from wind power? Possibly, but we just don't know: all low-carbon approaches face challenges that "might include lack of viable engineering pathways to substitutions."[211] On top of the elaborate technical challenges, we can add to that the political and trade complications of regulating industrial heat.[212]

Nor is this the end of infrastructure issues. If we are to get *all* our energy from renewable electricity, that requires a vast expansion of the electric grid. Our current electric grid is aging and already requires many billions of dollars in investments to maintain.[213] One researcher estimated that it would take five trillion dollars to replace the current grid.[214] If we go to 100 percent renewables and an all-electric economy, then the electric grid would be several times larger and would have to carry *almost all the energy* used in the United States. This would require money, energy, and materials to build and maintain— another factor chipping away at the EROEI of renewables.

210 Friedmann, S. J. et al., "Low-carbon Heat Solutions for Heavy Industry," pp. 8, 59, 60.
211 Friedmann, S. J. et al., "Low-carbon Heat Solutions for Heavy Industry," p. 59.
212 Roberts, "This climate problem is bigger than cars."
213 American Society of Civil Engineers, "2017 Report Card."
214 Rhodes, "The old, dirty, creaky US electric grid would cost $5 trillion to replace."

3. Materials issues: goodbye oil, hello neodymium

Solar power and wind turbines will obviously not be limited due to a shortage of sunshine or wind. But they could become limited because of a shortage of the metals used to create solar panels and wind turbines.

Certain advanced solar PV technologies are critically limited by shortages of the rare earth metals ruthenium, gallium, and indium.[215] Concentrated solar power (CSP) relies on silver, which is used because of its high reflectivity, but silver is limited. Some types of renewables, such as wind turbines, require neodymium for optimum efficiency. Neodymium is widely used in wind turbines; a high-performance wind turbine can require two tons of neodymium.[216]

"To provide most of our power through renewables would take hundreds of times the amount of rare earth metals that we are mining today," according to Thomas Graedel, a pioneer in the field of industrial ecology.[217] What's more, these rare earth minerals are also needed for laptop computers, smart phones, televisions, and hybrid cars, creating a conflict between renewables and much of today's modern technology, with all of it dependent on international trade as well as supplies of intrinsically rare materials.

None of these technical limitations are absolute. We could use other, more common substances—such as just silicon for solar panels and other (less reflective) substances than silver for our CSP reflectors—and adapt to less efficient wind turbines. All these efforts, however, will further limit EROEI.

....

215 Grandel and Höök, "Assessing Rare Metal Availability Challenges."
216 Jones, "A Scarcity of Rare Metals Is Hindering Green Technologies."
217 Cho, "Rare Earth Metals."

In addition to these specialized rare earth elements, renewables often use large amounts of common materials such as steel and concrete. In 2003, nuclear engineer Per F. Peterson estimated that a typical wind turbine required 460 megatons of steel and 870 cubic meters of concrete per average megawatt of generating capacity, while coal required 98 megatons of steel and 160 cubic meters of concrete.[218] In 2010, researchers at Argonne National Laboratory concluded that the mass-to-power ratios of concrete and steel needed for wind turbines "are roughly two to five times higher than those of conventional systems."[219] We can hope for increased efficiency, but we are still talking about a *huge* increase in the concrete and steel required.

4. Land use: the "footprint" of renewables

How much physical space do renewables take up? The area taken up by solar PV panels, in relation to the energy generated, is significant but doesn't pose an immediate limit on solar PV. Many solar panels are mounted on the roofs of buildings, which are often not used for anything else anyway.

Wind power is different. Wind turbines have relatively the best EROEI of renewables, but they also have a considerable "footprint." While the physical base of the wind turbine isn't large, wind turbines must be spaced away from each other or they will interfere with each other's ability to create electricity from wind. Within much of this footprint, there is theoretically room for other noncompetitive uses, such as farming. But wind would be extremely inconvenient for homes or businesses. Wind turbines are rather noisy, creating what many describe as a "swishing" or "thumping" sound, annoying not so much

....

218 Peterson, Per F. "Will the United States Need a Second Geologic Repository?"
219 Sullivan et al., "Life Cycle Analysis Results," p. 47.

because of the absolute loudness but because it extends into very low frequencies.[220]

On top of that, there are only so many suitably windy sites around. To power the entire United States on wind power might require an area of 500,000 square kilometers (an area greater than that of the state of California).[221] Because wind turbines pose a substantial footprint issue, we will need to think carefully about where we site them.[222]

5. Transition issues (the "renewables gap")

Building out a renewable infrastructure requires materials and energy. We would need to build wind turbines and solar panels, expand the electric grid, create electrified rail transport and electric cars, and provide for energy storage during those intermittent downtimes for renewable-energy sources.

Where is all this energy, and all these materials, going to come from? Energy supplies are already increasingly constrained due to peak oil issues. Renewable energy generally requires "up front" energy investment; while the net result is favorable, most of the energy spent on renewable infrastructure is spent before a single kilowatt hour is generated.

We could probably increase energy and materials production temporarily to a certain extent. But most likely, at least some—possibly all—of the energy and materials needed, would be diverted from the consumer economy. This may not be a popular move!

....

220 Expert Panel, "Understanding the Evidence: Wind Turbine Noise," p. 21.

221 Clack et al., "Evaluation of a proposal for reliable low-cost grid power," p. 5.

222 MacKay, *Sustainable Energy,* and Cobb, "How Many Windmills Does It Take to Power the World?"

The gap between the energy and materials needed to produce this infrastructure, and the energy and materials that are *politically possible* without denting the consumer economy, creates the "renewables gap."[223] The United States faced the same sort of problem during the Second World War; we couldn't have both "guns and butter." This problem was resolved after Pearl Harbor: the entire country mobilized, the domestic auto industry was shut down, many consumer items disappeared or were greatly reduced, and "victory gardens" provided about 40 percent of all vegetable production in the United States. Diverting resources from the consumer economy creates a *political* problem, not a technical problem. Awareness of the sacrifices that will be required hasn't really penetrated the public's consciousness, and it isn't clear how receptive the public will be.

Previous energy transitions in the United States have taken a long time under relatively favorable conditions. In 1776, the United States was almost completely dependent on wood, and it was about a century before more than half of the country's energy came from something *other* than wood. Later in the nineteenth century, another energy transition—from the coal era to the oil era—took many decades. Why would the transition from fossil fuels to renewables be any quicker?

Because the transition from fossil fuels to renewables is likely to involve substantial sacrifices on the part of the consumer economy, the *politically available* energy and materials may be less than what is needed. If time passes with no action on renewable infrastructure, and total available energy begins to decline due to oil depletion, the renewables gap will grow larger, thus increasing the chances that we will be caught short. The longer we wait, the more traumatic this transition will be, and the more traumatic it will be, the more resistance there will be.

....

223 Vail, "The Renewables Gap."

Financial costs can be evaded or postponed by printing money, but we can't print oil.

Other renewable-energy systems

Corn ethanol, energy-wise, has been a disaster in the United States: without some creative accounting, the EROEI of corn ethanol is actually less than 1:1,[224] and it *increases* greenhouse gas emissions (all that corn is grown with extensive use of fossil fuels).[225] There have been proposals to grow biomass for biofuels in a more efficient manner, perhaps using genetically engineered algae to create something like ethanol with much greater efficiency than traditional biomass, but so far this is still under development.

Hydroelectric power, as previously mentioned, is already limited. Moreover, the reservoirs behind the dams eventually silt up. We might then tear down the dams and rebuild them somewhere else in the same river system. However, dams have other adverse environmental impacts, such as the flooding of large areas of vegetated land, which destroys habitat and creates methane from the rotting drowned vegetation.

Traditional energy biomass, in the form of wood, is what powered the human economy for thousands of years. However, it long ago posed a limit to growth. Even with a much smaller population and lower standard of living, deforestation was rampant in Europe in the Middle Ages. By 1500, Europe faced shortages of wood for fuel.[226]

Within these limitations, though, biomass does fill a number of needs not met (or not easily met) by electricity. First, it is renewable. The world ran on wood biomass for thousands of

224 Pimentel and Patzek, "Ethanol Production Using Corn, Switchgrass, and Wood."
225 McMahon and Witting, "Corn ethanol and climate change."
226 Kwiatkowska, "The Sadness of the Woods Is Bright."

years, so we know it works. Most importantly, biomass and biodiesel (diesel fuel produced from plants) are able to generate the high heat needed for much manufacturing, and biodiesel is dense enough to keep trucks running, which is difficult for renewable electricity.[227]

But we need to keep in mind that if we are talking about biomass as a primary source of energy, we are talking about an economy *much* smaller than today's. The population of the world in 1500 was on the order of 500 million, less than 10 percent of today's population, and the standard of living for almost everyone (even kings and queens) was much lower than that of the average American. On the other hand, hydroelectric power could also continue to produce electricity. With five centuries of accumulated scientific knowledge, plus biomass energy and hydroelectric power, technically advanced civilizations with a much smaller human population may be able to continue and even thrive on renewable energy, no matter what problems we have with modern solar and wind.

Conclusions

Renewable energy is a key part of an overall approach to limits to growth. People are working on all these problems, but we are nowhere near having anything resembling a "national conversation" about what, in practical terms, the research, development, and implementation of such alternatives will require. Because there are so many plausible objections that reduce EROEI, intuitively it is doubtful that with current renewables technology we can maintain anything remotely resembling our modern consumer society with 100 percent renewable energy. Technology could be our friend concerning renewables, and we may be pleasantly surprised with new technological developments with batteries or power-to-gas

227 Friedemann, *Life After Fossil Fuels*, p. 85.

generation. It is doubtful, though, that we can *count* on this sort of thing.

The only truly renewable and sustainable energy system is likely to be one that produces much less energy. It is unlikely that an economy powered by renewables could ever produce the quantity of goods and services that our current economy does. Until someone can clearly show how such a gigantic renewable-energy economy could work in the United States and the rest of the world, we need to think about how renewables will work in a post-industrial world with a much smaller population and much smaller economy. This doesn't necessarily mean we shouldn't develop renewables anyway, just that we should be realistic about what they can achieve.

CHAPTER FIFTEEN

NUCLEAR POWER TO THE RESCUE?

The limits of atomic energy

Nuclear energy isn't renewable, as it is dependent on a limited supply of fissionable nuclear fuels such as uranium. Nuclear power creates radioactive waste products, such as the extremely toxic and long-lived plutonium. It is complex, disparaged by many environmentalists, and feared by the general public. Three widely publicized nuclear accidents (Three Mile Island, Chernobyl, and Fukushima) haven't enhanced its popularity. Nuclear weapons have the capability to wipe out the human race. Nuclear power plants are expensive and difficult to finance. The EROEI of nuclear power is said to be relatively low, estimated by some at about 5:1—comparable to that of the Alberta tar sands.[228]

Fusion power using *hydrogen* (rather than fission of uranium atoms) might provide practically limitless clean energy. But fusion must overcome epic technical obstacles, starting with the need to generate extraordinarily high temperatures (hotter

228 Inman, "How to Measure the True Cost of Fossil Fuels."

than those found in the core of the sun) and the need for super-powerful magnetic fields.[229] One thing to remember about fusion, as well, is that even if it becomes practical and commercial, fusion reactors will wear out and need replacement, which in turn will require scarce metals and minerals. Work on fusion continues[230] and though commercialization still seems many years off, there has been a notable expansion in investment in nuclear fusion research.[231] But problems with fusion over the past fifty years have prompted a long-standing quip: "Fusion is the energy source of the future—and it always will be."

So why are we even talking about nuclear power?

Nuclear overview

We are talking about nuclear because we face critical limits on energy and a climate crisis. Renewables *may* be the answer, but no one can say precisely how a rollout of a renewable infrastructure would go. If building out a renewable infrastructure turns out to be considerably more difficult than people are expecting—and it probably will—nuclear energy may look considerably more appealing than it currently does. While environmentalists have raised a wide array of plausible objections against nuclear power, it has attracted some environmental support. James Hansen, a scientist and a leader in the fight for climate change action; James Lovelock, the originator of the "Gaia" hypothesis; and British environmental activist George Monbiot all support nuclear power.

There are several areas where nuclear is more advantageous than renewables. As we have seen, renewables suffer from storage and intermittency issues, require heavy up-front energy investment, and have a relatively large physical footprint. Nuclear

229 Kimani, "The Holy Grail of Energy is Finally Within Reach."
230 Kimani, "The Holy Grail of Energy Is Finally Within Reach."
231 Reed, S., "Nuclear Fusion Edges Toward the Mainstream."

plants can be run continuously to provide *base-load electricity*, the minimum amount of electric power needed for an area over a twenty-four-hour period. They can be sited almost anywhere and have a relatively small physical footprint. Nuclear also has a much smaller metals-and-materials requirement per unit of energy generated; a 1970s-era nuclear power plant requires less than a quarter of the concrete and less than 10 percent of the steel that wind power does.[232] Molten salt reactors (discussed below) could provide high levels of industrial heat that much of modern heavy industry requires, thus potentially enjoying a key advantage over renewables. Like renewables, nuclear power has very few greenhouse gas emissions, mostly connected with the initial construction of the nuclear plant.

Problems with nuclear

That's the good news. But what about the problems with nuclear power?

1. **Nuclear power accidents.** There have been three infamous nuclear power accidents since the Second World War: Three Mile Island in the United States, Chernobyl in the former Soviet Union (now Ukraine), and Fukushima in Japan. The argument from nuclear advocates is that while nuclear is risky, it doesn't look quite so bad in terms of the *relative* risk (to both workers and the public) of various power sources.

Three Mile Island didn't generate any immediate deaths. Chernobyl resulted in forty-seven immediate deaths. Fukushima didn't generate any direct radiation deaths, although several dozen people had significant radiation exposure and about 1,600 died in the panic of the evacuation from the area (unrelated to

....

232 Peterson, "Will the United States Need a Second Geologic Repository?"

any direct exposure to radiation).[233]

What about *ultimate* deaths from Chernobyl and Fukushima, including cancer deaths from radiation exposure? The World Health Organization (WHO) estimated about four hundred ultimate cancer deaths for Fukushima; others estimated from four thousand to sixty thousand for Chernobyl.[234] Some made *very* large estimates for Fukushima, estimating that a million people might die.[235] The reality is that the Fukushima accident caused almost no deaths (if any) from radiation. Most people who died were killed in the panic of the evacuation.

Chernobyl was more serious: the accident caused 134 cases of acute radiation syndrome among the workers who tried to contain the damage, of whom forty-seven later died, though not all of them from radiation. The United Nations Scientific Committee on the Effects of Atomic Radiation (UNSCEAR) additionally found that there were 6,848 cases of thyroid cancer in young children, due to the Soviet Union's failure to prevent people from drinking milk containing radioactive iodine.[236]

None of this is good, but as a whole, evidence is weak for any large-scale health effects due to radiation from either the Chernobyl or Fukushima accidents.[237] The atomic attacks on Hiroshima and Nagasaki provide something of a benchmark for evaluating radiation deaths; the survivors of Hiroshima and Nagasaki have been well studied for many decades. There were 200,000 deaths from falling debris, flying glass, burns, etc., for those close to the center of the blasts; however, there were almost no cancer deaths due to radiation for anyone who was

233 Johnson, "When radiation isn't the real risk."
234 Ritchie, "What was the death toll from Chernobyl and Fukushima?"
235 Ji, "'Million Cancer Deaths from Fukushima Expected in Japan,' New Report Reveals."
236 Monbiot, "Evidence Meltdown."
237 Monbiot, "Evidence Meltdown."

more than about two miles away.[238] It is hard to imagine that Fukushima and Chernobyl created *more* danger to humans than the atomic attacks on Japan in 1945, and this is consonant with other studies on the effects of radiation.[239]

How do Chernobyl's casualties compare to those from coal mining? Mining accidents worldwide kill about 12,000 people a year,[240] many of them from mining coal. In China, annual fatalities from coal mining (which had been as high as 7,000 in 2002), only recently fell below 1,000 per year.[241] In the US, coal mining is relatively safe in terms of mining accidents.[242]

But this doesn't count black lung disease. The US reported more than 76,000 deaths from black lung disease since 1968.[243] Black lung experts have warned of a resurgence of the disease; autopsies done on miners killed in an explosion in West Virginia in 2010 indicated that more than two-thirds of them had black lung disease.[244] China reported more than half a million cases (and 140,000 deaths) from the 1950s through 2005.[245] Switching from coal to nuclear power would actually make energy production relatively safer in terms of risk of casualties per unit energy produced.[246]

....

238 Radiation Effects Research Foundation, "Frequently asked questions."

239 Kharecha and Hansen, "Prevented Mortality and Greenhouse Gas Emissions."

240 Lang, "The dangers of mining around the world."

241 Lelyveld, "China Cuts Coal Mine Deaths, But Count in Doubt."

242 United States Department of Labor, "Coal Fatalities for 1900 through 2018."

243 United States Department of Labor, "MSHA issues final rule."

244 Berkes, "As Mine Protections Fail, Black Lung Cases Surge."

245 People's Daily Online, "Black lung disease claims 140,000 lives in China."

246 Kharecha and Hansen, "Prevented Mortality and Greenhouse Gas Emissions."

2. **Nuclear waste.** In addition, the *normal* process of using nuclear power results in radioactive nuclear waste, which must be disposed of somehow—and which has been for decades a contentious political issue in the United States. The most prominent and dangerous nuclear waste is plutonium. The half-life of plutonium (Pu-239) is about 24,000 years, which is many times longer than any known human civilization has survived.

Nuclear advocates say that the waste can be safely buried somewhere (either deep in the sea or at Yucca Mountain in Nevada). However, in light of plutonium's long half-life, concerns exist as to whether whatever solution we devise will work for the next ten thousand years. If nuclear power goes mainstream and provides the whole world with energy for several centuries, even a small chance of some serious nuclear contamination would be multiplied considerably.

These problems are partially offset by two factors. First, the *physical* footprint of radioactive waste isn't that great. Nuclear advocates point out that a one-thousand-megawatt nuclear power plant (capable of supplying a large city with electricity) would create about thirty tons of high-level nuclear waste per year, but a similar coal plant would create 300,000 tons of ash.[247] Nuclear waste is highly toxic, but there's not much of it. Second, in *principle*, much nuclear waste doesn't need to be wasted. Using a molten salt reactor, it is possible to use unconsumed U-235, as well as U-238 (from "depleted" uranium), and any newly created plutonium as nuclear fuel. There are still radioactive waste products, but they are much less dangerous and long-lasting than plutonium.

3. **Proliferation of nuclear weapons and security.** Doesn't nuclear power make nuclear war more likely? Some of the basic techniques needed for nuclear power, such as mining and refining nuclear material, are similar to those needed to build a bomb.

247 World Nuclear Association, "What is Nuclear Waste?"

This probably makes the world somewhat less safe. But using nuclear power doesn't *increase* nuclear security risks by much; the genie is out of the bottle. Nuclear technology is well understood. India, Pakistan, Israel, and even North Korea have the bomb, despite non-proliferation treaties.

The biggest proliferation issue is the fear of unwanted parties acquiring plutonium. Neither natural uranium (less than 1 percent U-235) nor even enriched uranium (3–8 percent U-235) is much use in building a bomb, which requires more than 90 percent U-235. It also requires expensive and sophisticated enrichment facilities to separate the fissile U-235 from U-238; building such facilities is much more difficult than acquiring uranium in the first place.

Plutonium, however, *does* represent a proliferation problem. It is chemically different from uranium and therefore can be easily separated (no enrichment facilities are needed). That's why nuclear power for Iran was (and still is) such a contentious issue. Some have said that the Obama-era nuclear deal with Iran (from which President Trump withdrew in 2018) made it easier for Iran to get a nuclear weapon, while many of the agreement's provisions were intended to establish elaborate safeguards to make construction of a weapon much more difficult.[248]

However, while nuclear power plants *do* present some security risk, they don't significantly *increase* the proliferation security risk for countries (such as the United States, Russia, and China) that already have the bomb and that signed the 1968 Nuclear Non-Proliferation Treaty.

4. **Cost.** Why are nuclear power plants so costly to construct? Nuclear advocates say it is because nuclear power plants are held to much stricter standards than fossil fuel plants, and that if they were held to the same standards, nuclear would be cost-competitive. In the United States, it is more difficult

---

248 Broad and Peçanha, "The Iran Nuclear Deal – A Simple Guide."

to secure approval of new and safer nuclear designs from the Nuclear Regulatory Commission than to get the older designs approved simply because the older (but riskier) designs are the ones with which we have experience.[249] In the current political environment, no one wants to advocate tighter standards for coal plants—however much that might make sense—due to the power of the coal industry and inevitable complaints from consumers about higher prices. Nor is anyone likely to advocate looser standards for nuclear power after Fukushima. So we are stuck with coal rather than nuclear.

5. **Nuclear fuel supplies.** In the long run, this is the fundamental stumbling block for nuclear power. Just like fossil fuels, when nuclear fuels are gone, they're gone.

Uranium isn't renewable. One source has stated that we have already reached a worldwide upper production limit for uranium extraction and that production will soon decline.[250] However, nuclear supply may be several hundred times larger than the supply of U-235 (the isotope of uranium that can be fissioned for energy). The vastly more plentiful U-238 can be turned into plutonium and used as a nuclear fuel. In addition, thorium—which is three times as plentiful in the earth's crust as *all* isotopes of uranium—can also be turned into U-233 and used as a nuclear fuel. On the other hand, even this expanded supply of nuclear fuel won't last forever, especially if we expand nuclear power to supply almost all the world's energy needs.

There is also active discussion about extracting uranium from seawater, where there is a vast and virtually infinite supply. But it seems somewhat doubtful that the EROEI for this process is sufficiently great to make it energetically worthwhile.[251]

249 Batkins, Sam et al., "Putting Nuclear Regulatory Costs in Context."
250 Dittmar, "The End of Cheap Uranium."
251 Bardi, "Extracting Minerals from Seawater."

Box 15-1. Thorium versus uranium fuel cycle		
To get 1,000 megawatts of electric energy for one year:		
Step	**Uranium cycle**	**Thorium cycle**
Mining	800,000 tons of ore	200 tons of ore
Ore extraction	250 tons of natural uranium	1 ton of natural thorium
Ore refinement	End result: 35 tons of "enriched" uranium (~1 ton U-235 and ~34 tons U-238) *plus* 215 tons of "depleted" uranium (U-238) *Very expensive process*	No refinement needed
Nuclear fuel processing ("burning")	Uranium-235 content is "burned" out of the fuel; some plutonium is formed and burned.	Thorium is introduced into blanket of fluoride reactor, completely converted to U-233, and "burned."
Waste products	35 tons of spent fuel (Yucca Mountain for 10,000 years): 33.4 tons U-238 0.30 tons U-235 0.30 tons plutonium 1 ton of other fission products	1 ton of fission products: (a) 0.87 tons decays within 10 years to products with background radiation; some products can be sold (b) 0.13 tons must be isolated for 300 years
Source: Bonometti, "The Liquid Fluoride Thorium Reactor."		

6. **EROEI (energy return on energy invested).** Advocates and detractors of nuclear power give wildly different estimates of the EROEI of nuclear power. Some people say it is close to 5:1.[252] But others say it is more like 50:1 or 75:1, more akin to hydroelectric power.[253] Controversy surrounds the energy costs of uranium enrichment, nuclear accident cleanup, and nuclear waste disposal. In the early days of nuclear power, uranium enrichment was its main energy cost, which was a key factor in calculating the lower EROEI figure, but this would likely not hold true today (and wouldn't apply at all if thorium rather than uranium were used).

Estimating the *energy* cost of cleaning up the Fukushima accident, or future accidents that may or may not happen, enmeshes us in further imponderables. It seems likely that the EROEI of nuclear is substantially higher than 5:1, but a realistic and objective figure depends on many technical variables.

7. **Electricity.** This is one disadvantage that nuclear power shares with renewables: the primary result is energy in the form of electricity, which doesn't work well in some uses for transportation, mining, and manufacturing. Industrial heat is particularly problematic. Molten salt reactors (see below) can generate temperatures up to 700 degrees Celsius, which will satisfy some but not all industrial heat requirements.

8. **Complexity.** Complexity may be a problem, depending on who needs to understand the process and why. Nonexperts and the general public often find it difficult to assess the sheer technical complexity surrounding nuclear power. There are multiple nuclear fuels that can be used and a variety of reactor designs. Some of nuclear power's technical complexities aren't just engineering matters; they also matter for public concerns

252 This is the estimate given by Inman, "How to Measure the True Cost of Fossil Fuels."
253 Mearns, "ERoEI for beginners."

about safety issues. (Wind turbine technology is also complex, but the complexity doesn't obviously affect public safety.) Complexity thus makes nuclear power a political problem vulnerable to public criticism. Faced with a barrage of technical information, much of the public will likely just throw up its hands and lump all nuclear power plants together with Chernobyl and Fukushima. Complexity may also cause accidents directly; the accidents at Three Mile Island and Chernobyl were brought about or exacerbated by a combination of people who didn't have the expertise or training to know what they were doing.

Molten salt reactors and thorium

In the 1960s, the technology existed to significantly circumvent three key areas of concern about nuclear power: wastes, accidents, and supplies. This was the molten salt reactor based on thorium that was developed in Oak Ridge, Tennessee, and ran successfully for four years. (In 2021, China announced plans to test a molten salt reactor using thorium, the first one since the Oak Ridge reactor.[254]) This broaches more technical issues that are too lengthy to fully discuss, but there are two significant differences between the Oak Ridge reactor and most nuclear reactors. (See Box 15-1, "Thorium versus uranium fuel cycle.")

1. *The thorium cycle is safer than the natural uranium cycle.* Thorium is more abundant than uranium, the radioactive waste products are fewer and less dangerous, the uranium "enrichment process" is unnecessary, and the use of plutonium is avoided completely.

2. *Molten salt reactors are inherently safer than light water reactors (and most other reactor designs).* Light water reactors depend on continuous pumping of water as a coolant for

254 Mallapaty, "China prepares to test thorium-fuelled nuclear reactor."

solid nuclear fuel to prevent it from overheating and melting down, and that water must be maintained at high pressures much greater than atmospheric pressure. If the water supply is interrupted, the core will overheat and a meltdown of the reactor core can occur, as happened at both Three Mile Island and Fukushima.

Molten salt reactors use molten salt as a medium rather than water. The nuclear fuel is dissolved in the liquid molten salt and circulates within it. It can't melt down because it's already melted, and it doesn't require pressurization so can operate at standard atmospheric pressure. The design is "passively safe": if the reactor overheats due to a power failure, a "freeze plug" will melt, the molten salts will drain out into a large containment vessel, the fluid will expand, and a nuclear reaction becomes impossible. In principle, molten salt reactors could also use nuclear wastes (depleted uranium and unused plutonium) as fuel, thus simultaneously solving much of the nuclear waste problem and providing a much-expanded source of nuclear fuel.[255] Molten salt reactors could also provide a low-carbon source of industrial heat, which is very expensive to do using electricity. Nuclear power can produce temperatures of 700 degrees Celsius;[256] temperatures of 550 degrees Celsius would supply more than half of the industrial heat market.[257] However, we don't know whether higher temperatures would be possible, and so far no one has actually used nuclear power for industrial heat.[258]

255 Williams, Stephen, "How Molten Salt Reactors Might Spell a Nuclear Energy Revolution."
256 Waldrop, "Nuclear goes retro—with a much greener outlook."
257 Thorium MSR Foundation, "In Depth: Clean Industrial Heat."
258 Friedmann, S. J., "Low-carbon Heat Solutions for Heavy Industry," p. 23.

Conclusions

Renewables enjoy much more public confidence than nuclear power. We have an indefinite supply of wind and sun. Renewable technology is easier to understand and can't be easily weaponized. Because of this, renewables have become the default public policy choice of those concerned about alternatives to fossil fuels in our energy future.

But none of the common objections to nuclear power automatically exclude nuclear power from consideration. If the transition to a renewable-energy economy stumbles badly—and there seems to be a considerable chance this will happen— nuclear energy may not look quite as bad as it does to many people today. We may face a choice not between renewables or nuclear, but between fossil fuels or nuclear. We need to understand the issues before we are suddenly faced with such a dilemma.

Nuclear does provide some advantages over wind and solar in several respects: (1) nuclear largely solves many intermittency problems; (2) nuclear can be sited anywhere (it doesn't depend on finding windy or sunny spots); (3) nuclear has a much smaller physical footprint; (4) nuclear has a smaller metals-and-minerals requirement per unit energy; and (5) nuclear could provide at least some industrial heat more easily.

The fundamental problem with nuclear power is that nuclear fuels are finite, just like fossil fuels, and so provide at best a temporary solution to energy problems. But how temporary? Nuclear advocates typically talk in terms that suggest continuing the current growth economy, only avoiding greenhouse gas emissions in the process. As far as we know, no one is talking about nuclear power in the context of a smaller degrowth economy. How long could a smaller economy continue, using a finite supply of nuclear material, and with what level of energy consumption? Can we rule out the possibility that thorium and molten salt reactors might power a much smaller economy for thousands of years? This isn't forever, but it's a long time. Because nuclear power is currently *persona non grata* to most of the environmental movement, no one has bothered to investigate these questions. Until we have better information, we don't know.

Neither renewables nor nuclear power, by themselves or together, can avert *ultimate* limits to growth, simply because we face many limits other than limits on energy. Energy can help, but it can't offset limits on other resources indefinitely.

CHAPTER SIXTEEN

HALF-EARTH

A place for wilderness on the planet

The key element in embracing limits is adopting a new relationship with nature. The most difficult environmental issue humans must face isn't energy; it's biology. The bloated biomass of humans and their livestock is crowding out all other living things on the planet, undercutting the basis of the natural world that supports human existence. We need to do more than just protect existing wilderness areas; we need to expand them, and a number of people and groups have suggested ways of doing that in dramatic fashion.

The most radical of these is the distinguished American naturalist E. O. Wilson's book *Half-Earth*.[259] Wilson proposes setting aside half of the planet's land area for wilderness. Prior to Wilson, others had made proposals in the same vein. Deborah and Frank Popper, a geographer and engineer husband-and-wife team, proposed to return most of the American Great Plains to wilderness, which they termed "the Buffalo Commons." Geoscientist Paul Martin and his associates suggested taking

---

259 E. O. Wilson, *Half-Earth*.

wilderness back not just to pre-industrial times but to prehistoric times, with "Pleistocene rewilding." A growing animal rights movement seeks to end the captivity of billions and billions of domesticated animals who suffer horribly. All these people recognize a central issue: we need to dramatically reshape our behavior toward the natural world.

E. O. Wilson

E. O. Wilson (Edward O. Wilson), the famed American biologist, researcher, theorist, naturalist, and author, proposes "committing half of the planet's surface to nature."[260] There's an obvious question here: What does this actually mean? Which half of the planet does he propose be set aside for wildlife?

And what about the oceans? Wilson's formulation refers only to the *land area* of the planet, but *Half-Earth* does include a brief chapter on marine biodiversity. He notes overfishing, the death of coral reefs, and a rapidly expanding human footprint in the oceans; but despite all this, ocean biodiversity isn't quite as threatened as land biodiversity.[261] The total biomass of fish (which still surpasses that of humans and livestock) has decreased by about half under human influence,[262] though the biomass of marine *mammals* (e.g., whales and dolphins) has declined much more steeply, by about 80 percent.[263] In the case of the oceans, obviously the key means to restrict the human footprint is to restrict fishing.

260 Wilson, *Half-Earth*, p. 3.
261 Wilson, *Half-Earth*, p. 114: "Yet most of marine biodiversity persists."
262 Bar-On et al., "The biomass distribution on Earth," Supplementary Information Index, p. 60: before human fishing, the biomass of fisheries was approximately twice as large.
263 Bar-On et al., "The biomass distribution on Earth," p. 6508: human exploitation has led to "approximately fivefold decrease in marine mammal global biomass."

While Wilson is eloquent and knowledgeable in his defense of biodiversity, he is frustratingly coy about exactly how his proposal would work. He contents himself with discussing the immensity and complexity of the wild world and gives some examples of ecosystems he wants to preserve. But in *Half-Earth*, he doesn't explicitly lay out an overall plan. It may be that he doesn't know himself how it will work, or that he knows but is afraid to commit it to writing. Either way, it is left for us to spell out the necessary conclusion: if "half-earth" is to be meaningful, that means eliminating or drastically reducing the livestock industry. We can't really begin to discuss the problem of restoring marine diversity, except to say it should be obvious that this, too, requires at the least sizable restrictions on the fishing industry.

To be meaningful, nature's "half" of the planet's land areas must include biologically productive land. If we start by allocating all the desolate areas of the planet to "wilderness"— e.g., Greenland, Antarctica, the Sahara Desert—we won't be doing wilderness or wildlife any significant favors, although perhaps the penguins in Antarctica will appreciate the gesture.

One easy-to-formulate plan would start with livestock agriculture, which is the biggest human use of land on the planet. Livestock agriculture is the bulk of our biomass problem in the first place. Eliminate the livestock industry and allocate those lands (grazing land and croplands growing food for livestock) to wilderness. This doesn't displace any significant number of humans from space we now inhabit, except perhaps the occasional cowboy camping out underneath the stars.

It's not going to be *quite* that simple. First, farmers, ranchers, and others who derive their income from this land would need some alternative means of employment or support. This is a problem shared with others currently employed in ecologically destructive industries, such as coal miners and oil-rig operators, who would be unemployed by policies protecting the environment (discussed in more detail in chapter 21).

Second, even if we could persuade most of the world to adopt a plant-based diet, much land has been degraded by humans. We'd need to undertake considerable restoration efforts in addition to simply abandoning the use of livestock. We'd need to connect wilderness areas for migratory species and take steps to defend and protect specific ecosystems, perhaps along the lines of our existing national parks and wilderness areas.[264] This is a breathtaking undertaking itself that we can't begin to address within the scope of this book.

Just eliminating most or all livestock agriculture would take us a considerable distance down the path we need to go to reach the goal of reserving half of the earth for wilderness. By allocating lands now used for livestock to wilderness, we could return about one-third of the biologically productive areas of the planet to wilderness. This would be comprised mostly of grazing land but also quite a bit of cropland. As an interim measure, just giving grazing land back to wilderness—which would leave much meat consumption in place, since much livestock (e.g., poultry, pork) doesn't use ruminant or grazing animals—would advance the goals of both reforestation and biodiversity.

This doesn't completely solve the biodiversity crisis. We also need to reduce human population numbers, over the coming centuries, to numbers closer to pre-industrial levels. But this plan displaces few if any humans from their homes; without even beginning to reduce the bloated human population, we will have made significant progress toward allocating close to half of the biologically productive land area on the planet to nature.

---

264 Wilson, *Half-Earth*, chapter 15.

The Buffalo Commons

As noted above, Wilson wasn't the first to propose returning human-occupied land to wilderness. In a 1987 article, Deborah and Frank Popper proposed returning much of the American Great Plains to wilderness.[265] The Poppers observed that much of the Great Plains is more densely populated than the region's resources can sustain. From a resource-management point of view, it would make sense to return the entire area to its "pre-white" state—basically, abandoned to wildlife, symbolically to the buffalo, thus the name "the Buffalo Commons." The presence of indigenous people in the area was insignificant in terms of ecological impact, and the Poppers proposed returning lands such as those in the Black Hills to the Sioux and other tribes.

The decline of the Great Plains ("decline," that is, in human terms) is not only inevitable but already underway: people are moving out. Many areas lack retail stores, interstate highways, and large hospitals. Vast swaths of land in the western United States, including most of Montana, Wyoming, North and South Dakota, and parts of Colorado, New Mexico, western Kansas, western Nebraska, and western Texas, qualify as "frontier counties"—defined as counties with fewer than seven residents per square mile. Occasionally one can find offers of "free land" in the Great Plains in exchange for living there or setting up a business for some minimal period of time. As of 2019, Flagler, Colorado (about 110 miles east of Denver), offered free land to businesses, while Marquette, Kansas (about 200 miles west of Kansas City), offered free land to anyone wanting to build a home there.

It's only possible to continue to prop up the rural economy through government subsidies or extraction of nonrenewable resources. Oil drilling (thanks to the recent rise of fracking for oil) has led to population increases in some areas, but the local

265 Popper and Popper, "The Great Plains: From Dust to Dust."

economies fall again when oil and natural gas prices sink or wells are depleted. In the meantime, soil erosion and groundwater depletion are accelerating. When humans first tried to cultivate the Great Plains early in the twentieth century, we created the Dust Bowl by plowing up deep-rooted perennial grasses. We have brought the area back by draining the Ogallala Aquifer (as discussed in chapter 9), but this strategy can't work forever. The Ogallala is finite, and in some areas in Texas it is already gone.

It is a false economy to continue to destroy the environment and deprive the local flora and fauna of habitat, simply to preserve this human environment. Instead of struggling against the inevitable, we should make the best of a bad deal by giving over the area to wilderness. The federal government could simply deprivatize much of the area and stock it with native species, including the buffalo. Introducing or re-introducing native species wouldn't be a simple task, but the first step would be to abandon human use of the area.

Many residents of the Great Plains were initially outraged by the Buffalo Commons idea, but as the decline of many rural counties has continued, many have grudgingly accepted or even embraced it.[266] Those who really want to stay could be integrated into the new wilderness economy, perhaps as tour guides to the new wilderness or as residents of retirement communities with fewer needs than those of most cities. As the Poppers wrote, "The small cities of the Plains will amount to urban islands in a shortgrass sea. The Buffalo Commons will become the world's largest historic preservation project, the ultimate national park."[267]

----

266 Williams, Florence, "Plains Sense."
267 Popper and Popper, "The Great Plains," p. 18.

Pleistocene rewilding

The "Buffalo Commons" and "Half-Earth" proposals strive to return land back to its state before the arrival of modern ("white") human civilization. But "Pleistocene rewilding," advocated by Paul Martin, Josh Donlan, and others, seeks to revert the land back to its *pre-human* state.

The Pleistocene is the most recent geological epoch, part of the Quaternary period, and lasted from about 2.6 million years ago to about 11,700 years ago. The *current* spate of human-caused extinctions began a bit before the Pleistocene ended, about thirty to fifty thousand years ago. The purpose of Pleistocene rewilding is to determine not where wilderness area should be but what "wilderness" is in the first place. To restore nature, we should return to something like the *status quo ante*—the environment *before* the beginning of the current geological epoch. Rewilding advocates argue for this on numerous grounds, including ecological, evolutionary, economic, aesthetic, and ethical.[268]

Primitive humans, before the widespread adoption of agriculture, hunted and (directly or indirectly) caused the extinction of such large animals as the mammoths, mastodons, giant sloths, and saber-tooth tigers. More than 70 percent of large animal species were wiped out in Australia, North America, and South America. Present-day flora and fauna evolved under the heavy-handed influence of humans, and this distorted not only the evolution of the species that went extinct, but also all the plant and animal species that remained.

While we can't yet "resurrect" extinct species such as the mammoths, in many cases we have close analogs or relatives of extinct species (see Box 16-1, "Extinct North American species and their endangered proxies"). We have elephants instead of wooly mammoths, who diverged from mammoths about six million years ago. The musk ox has been re-introduced in remote

268 Donlan et al., "Re-wilding North America."

parts of Russia.[269] The American horse, now extinct, was already accidentally rewilded when the European colonists imported their cousins from the Old World. Perhaps these horses are so difficult to "eradicate" because they were originally indigenous to the area anyway. Many of these modern-day relatives, like the elephant, are now themselves in danger of extinction; thus, Pleistocene rewilding would save nonextinct species as well as restore Pleistocene ecosystems.

Box 16-1. Extinct North American species and their endangered proxies	
Extinct or critically endangered species	*Proxy for rewilding*
American cheetah (*Acinonyx trumani*)	African cheetahs (*Acinonyx jubatus*)
Bolson tortoise (*Gopherus flavomarginatus*)	Not extinct (*northern Mexico*)
American horses	Feral horses (*Equus caballus*), Przewalski's horse (*E. przewalskii*), feral asses (*E. asinus*), Asian asses (*E. hemionus*)
American camel (*Camelops*)	Bactrian camels (*Camelus bactrianus*)
Proboscidians (mammoths, mastadons, gomphotheres)	Asian elephants (*elephas maximus*), African elephants (*loxodonta africana*)
American lion (*panthera leo atrox*)	Lion (*panthera leo*)
Source: based on Donlan et al., "Re-wilding North America."	

---

269 Anderson, "Welcome to Pleistocene Park."

The main criticism of or modification proposed to Pleistocene rewilding is that in the twelve thousand years since the end of the Pleistocene, evolution has gone further down the road, with species now adapting to the new circumstances we have created. It may not be possible to fully recreate a reasonable facsimile of the Pleistocene. Don't we have an obligation to members of these new species, forced to adapt through no fault of their own?[270] Older "rewilded" species may function, in effect, as invasive species to the new ecological mix that has evolved over the past twelve thousand years. The best solution in a given area might be a hybrid of new and old environments.

Rewilding isn't something we could accomplish overnight, but it does represent a new direction for environmental preservation of species, a direction that would not only slow down alteration of the environment to a more natural pace that would allow species to evolve and adapt more slowly, but also undo some of the damage.

Liberating domesticated animals

Domesticated animals, not wild animals, form the overwhelming bulk of large-animal biomass on the planet, but they are also the animals that suffer the most. Because of this, animal rights campaigners tend to focus on domesticated animals such as farmed animals and laboratory animals.

Many have written of the horrors of being a farm animal.[271] Animals are typically confined to very small spaces. The close confinement means that most animals raised for food are forced to live in, or close to, their own excrement and that of their

270 Marris, *Wild Souls*, makes this argument, and in chapter 8 (pp. 122–137) she suggests that there is no ethical obligation to maintain biodiversity because species aren't sentient.
271 See, e.g., Peter Singer, *Animal Liberation*; Matthew Scully, *Dominion*; and Jonathan Safran Foer, *Eating Animals*.

companions. To discourage the aggressive behaviors resulting from this overcrowding, they are also typically mutilated without anesthesia: tails docked, beaks removed, castrated, dehorned. Close confinement also creates a breeding ground for disease, so animals are continually dosed with antibiotics.

Most food animals have been deliberately bred to gain weight rapidly—or alternatively, to give huge and unnatural quantities of milk or lay unnaturally large numbers of eggs. This genetic manipulation often leads to health problems. Chickens become so heavy that their legs break, and dairy cows develop mastitis in their udders. And no food animal lives very long. When they grow up, they stop gaining weight and are slaughtered. Dairy cows and laying hens live somewhat longer, but they are sent to slaughter as soon as they cease being "productive." All of them die at an age equivalent to human teenagers.

What would "animal liberation" actually look like? The only true way to permanently "liberate" domesticated animals would be to return them to the wild, but many domesticated animals are unfit to exist in a truly wild situation. Cows, chickens, and pigs have been bred for odd properties that don't necessarily contribute to their survival in the wild, such as rapid weight gain and excessive egg-laying or milk production.

The world isn't likely to go vegan overnight. But if it did, the most humane thing to do with domesticated animals might be to spay or neuter most of them—those animals that couldn't survive and reproduce in the wild (domesticated turkeys), as well as those that might survive all too well as invasive species (feral pigs in Hawaii). It should be possible to re-introduce the wild ancestors of these species (or reasonable proxies) into the wild, however.

Why wilderness?

Establishing a true wilderness sphere means that, at least at first, humans will have to "police" nature—mostly, but not entirely, policing themselves. To a certain extent, this already happens now, for example with park rangers who defend large animals from poachers. This would have obvious benefits for all other species of plants and animals. But human "reverse engineering" of nature would also serve three valuable *human* purposes.

1. *Biodiversity*. These new wilderness areas could serve as wildlife habitat. Because a primary cause of species extinction is habitat loss, this could significantly reverse the biodiversity crisis in a relatively short time frame. To be effective the wilderness areas would need to include many varied habitats, including biologically productive areas previously seized for human purposes; otherwise, we would be prioritizing and giving an advantage to wild species in specific ecosystems.

2. *Climate*. Reforestation of wilderness areas would help reverse much of the carbon added to the atmosphere since the Industrial Revolution began (see chapter 7). Forests are a major carbon sink. We can't be sure exactly what share of the climate problem is caused by human disruption of natural habitats, but it is likely quite ample—maybe accounting for half of the total problem.

3. *Soil preservation*. Agriculture favors plants useful to humans (mostly, as food) by channeling plant growth in a particular direction. No matter how carefully it is practiced, it inevitably depletes the natural environment to an extent through soil erosion and soil depletion. Land taken out of agricultural use could function as "cropland reserves." As depleted soil is taken out of circulation (and reverted to wilderness), these soils would gradually recover (see chapter 8). The retired croplands could be replaced by healthy lands previously held aside as wilderness. Lands thus "rotated" to wilderness wouldn't qualify as true

wilderness for wild species, but they might serve as the next best thing and eventually acquire the same character as wilderness.

"Half-Earth" is the most radical way to make a *sizable* favorable impact on biodiversity and climate problems. Newly designated wilderness areas would instantly and dramatically expand habitat for wild animals, plants, and other living things. If we drew heavily from currently sparsely populated areas such as livestock grazing land, it would displace few humans from their habitat and also have health benefits, as we will see in the next chapter. This wouldn't be a quick and easy step, and even if taken wouldn't solve all our environmental problems, but it would be a very significant first step.

Most importantly (unlike with renewable energy or nuclear power), from the human end, there are no fundamental *technical* challenges to getting to Half-Earth. We know how to grow plants, and the infrastructure to do so is already there. The key obstacles are social and political. Such a shift is practically unthinkable in today's society.

Social and psychological consequences of "Half-Earth"

A worldwide "Half-Earth" policy would entail an immense cultural shift all around the world. To many people, even to many vegans and vegetarians, greatly reducing or eliminating livestock production must seem like a staggeringly difficult task. Even though a large-scale dietary shift to plant-based diets would be technically possible, there could be tremendous psychological resistance. Is it possible to overcome such psychological resistance? Unfortunately, this is a vitally important question we can't fully answer, although an increasing variety of interesting evidence bears on the subject and suggests that it is possible.

Our only recent experience with such large-scale shifts has come in times of necessity; during the First and Second World Wars, people in many parts of Europe shifted strongly toward vegetarian diets out of economic necessity (see chapter 17 below). But when the war ended, the people returned to their meat-eating habits.

Most people, without thinking, believe that meat consumption is "natural," "normal," "necessary," and "nice"—the "4 Ns" cited by social psychologists as the beliefs that underlie the largely unspoken ideology that favors meat consumption.[272] It must certainly *seem* that desire for meat is almost universal and transcends national, political, and class boundaries. Humans have eaten meat for thousands of years. In the modern world, when people experience a rise in their incomes, it frequently translates to an increase in their meat consumption.

But is this a *learned* behavior, deriving from culture and tradition, or does it have a biological or genetic basis? There is evidence that this is not biological or innate. In fact, the opposite is true; there seems to be a natural human affinity with all living things.

1. *The "meat paradox."* Most people believe they aren't cruel to animals, yet they know meat comes from an animal that has been killed. This is known as "the meat paradox," and it is now getting serious scientific attention. "This theoretical 'meat paradox' is a form of cognitive dissonance and has grave negative consequences for animal welfare and the environment," concludes one comprehensive literature review on the subject. The scientific literature supports the idea that the meat paradox is real and that eating meat triggers *cognitive dissonance*—the discomfort experienced as a result of a contradiction between one's beliefs and one's actions. The review also identifies a

272 Joy, *Why We Love Dogs*; Piazza et al., "Rationalizing Meat Consumption"; Zaraska, *Meathooked*.

number of possible "triggers" for this cognitive dissonance, such as being reminded of meat's animal origins or animal suffering, or sometimes just the presence of vegetarians. People may cope with this dissonance (without changing their eating habits) by, for example, minimizing animal suffering ("they don't really know what's happening"), asserting that humans are superior to animals, or saying that meat-eating is natural.[273]

Anecdotal evidence supports the idea of the meat paradox. Animal advocates often make exposés of brutal conditions on factory farms with graphic pictures of violence toward animals, while the livestock industry seeks to make such picture-taking illegal[274]—with both sides evidently assuming that violence toward animals is repulsive. Slaughterhouse workers, as well as animal-shelter employees charged with euthanizing "surplus" animals, often suffer psychological problems.[275] Many people view vegetarians as a threat to their cultural values, even though in objective terms vegetarians' beliefs promote human health and don't harm animals.[276] Denigrating vegetarians as "angry," "dogmatic," or "obnoxious" is likely another manifestation of cognitive dissonance.

2. *Biophilia.* Humans naturally love living things. Not only did E. O. Wilson make his Half-Earth proposal, but he also popularized the term *biophilia*, defined as the love of life and everything alive: animals, plants, and presumably fungi as

---

273 Gradidge et al., "A Structured Literature Review of the Meat Paradox."

274 Karlin, "Animal Slaughter Industry Making It Illegal to Show You Cruelty."

275 MacNair, *Perpetration-Induced Traumatic Stress*, pp. 87–88 (animal shelter employees, slaughterhouse workers). See also Eisnitz, *Slaughterhouse*; and McWilliams, "PTSD in the Slaughterhouse."

276 MacInnis and Hodson, "It Ain't Easy Eating Greens."

well.[277] Some argue that humans have a fundamental need for the natural world[278] and that children (and adults) can be harmed by "nature-deficit disorder."[279] With the disappearance of the natural world, the national parks are increasingly popular and increasingly crowded. Obviously, if we devote half the planet to nature, there will be a lot more of it to go around.

At least before culture and tradition have a chance to teach them otherwise, children have an instinctive affinity with animals. One recent study of children's attitudes toward animals concluded: "This study confirms that children are intrinsically motivated to treat animals well and to respect animal life."[280] Another study found that very young children often don't realize they are eating meat. The overwhelming majority of younger children classified food from cows and pigs as "not OK" to eat, but often thought animal foods came from plants. The study concluded, "It seems that children, especially preschoolers, assume that animals are not eaten."[281]

Animals feature prominently in children's stories and popular cartoons. While animals are sometimes presented in children's literature as threatening or dangerous, most popular stories portray animals as friendly or sympathetic. In Beatrix Potter's classic *The Tale of Peter Rabbit*, it is the *human* gardener Mr. McGregor who is presented as the comic villain who wants to put Peter Rabbit in a pie, while Peter is depicted in a sympathetic fashion. Peter wins out as the hero of the story, even though Mr. McGregor is simply trying to defend his garden vegetables against the rabbits and Peter Rabbit disobeys his mother by going into the garden in the first place. The popularity of this tale illustrates that children may feel a closer kinship with animals

....

277 Wilson, *Biophilia*.
278 Jones, *Losing Eden*.
279 Louv, *Last Child in the Woods*.
280 Fonseca et al., "Children's attitudes toward animals."
281 Hahn et al., "Children are unsuspecting meat eaters."

than with the rather mysterious and distant adult world, as Freud suggested.[282] Even though most adults in Western countries seem to adapt to knowledge (at some level) of such violence, the need to suppress that knowledge is bound to leave many people with some sort of psychological scar.

This is hardly conclusive proof, and this is an area of social psychology that has only recently come under serious scientific scrutiny. However, these different strands of evidence all suggest that instinctive revulsion against violence, as well as affinity with both plants and animals, is part of our biological make-up. Correspondingly, we might expect psychological and social benefits from the expansion of the natural world and more universal acceptance of plant-based diets.

Culture and tradition have a strong hold on our eating habits. For centuries, vegetarians have striven to change these habits, and many modern vegetarians and vegans devote quite a bit of energy and thought into how to accomplish this.[283] For the sake of the planet, we need these efforts to succeed.

Conclusions

Giving half of the planet to wilderness, which *still* leaves humans as the overwhelmingly dominant species, is a rational limit on human expansion. If we follow Wilson's advice on biodiversity, the two halves of the earth (wild and human-dominated) need to be equal in some sense. The only practical way to accomplish this balance requires sharply reducing or eliminating livestock agriculture, which could address multiple environmental problems at the same time.

....

282 Freud, *Totem and Taboo*, chapter 4.
283 See especially Leenaert, *How to Create a Vegan World*.

CHAPTER SEVENTEEN

FOOD, ANIMALS, AND DISEASE
The economics of nutrition

If we want to eliminate the livestock industry, we need to know about the practical effects on human health from avoiding animal foods. In many areas of human life, "embracing limits" means learning to live with less. In the case of food, it's likely that embracing limits means *better* health and a *higher* standard of living. That's why, if we're facing limits to growth, this is one "limit" we need to insist on.

There have been three great "epidemiological transitions" in the history of human diseases, through which human society moved from a state dominated by one set of health problems and longevity patterns to a different state with different problems and patterns.[284] All of them are connected with animals.

1. The first such transition began in prehistoric times after the development of agriculture. During this phase, the main health threat besides hunger was infectious disease—picked up from newly domesticated farm animals. Wild animals were much less

---

284 Harper and Armelagos, "The Changing Disease-Scape in the Third Epidemiological Transition."

a threat by this time. In general, the food supply and population expanded, although there was significant social inequality and food security was often something only elites could afford.

2. The second transition began in the twentieth century with the development of modern sanitation, antibiotics, and modern medicine. During this phase, infectious diseases receded as a problem, people lived longer, and the main health concern was chronic *degenerative* diseases such as heart disease and cancer— many of which are a consequence of *eating* animals.

3. The third phase started during the 1970s; while degenerative diseases continued to be a problem, new infectious diseases made a comeback. These emerging infectious diseases are also related to our treatment of or close contact with animals. COVID-19 and AIDS are the most spectacular examples in recent history, but they aren't unique.

In this chapter we will be concerned mostly with public health relating to the second and third phases above, both related to our treatment of animals: chronic degenerative diseases of overconsumption, and epidemics and pandemics from emerging infectious diseases.

It is worth reiterating that the first phase, the phase of infectious diseases, was *also* related to our treatment of animals. Most of these infectious diseases were picked up from animals thousands of years ago, as humans domesticated them and lived in proximity with them. These infectious diseases were the primary threat to human health during most of recorded history. We acquired tuberculosis from goats, measles from cows, whooping cough from pigs, typhoid fever from chickens, and leprosy from the water buffalo.[285] Indigenous people in the New World, lacking these livestock animals and with no exposure to these diseases, died in epidemics after the arrival of

285 See discussion in Greger, *Bird Flu*, pp. 89–90.

European colonists, their livestock, and their diseases, famously discussed in Jared Diamond's book *Guns, Germs, and Steel.*

Food and chronic degenerative diseases

The advent of modern medicine, vaccines, and modern sanitation practices has eliminated most of the threats from the infectious diseases that afflicted humans before the twentieth century. But with the rise of modern agriculture and the vastly increased consumption of animal products and refined foods, a new set of diseases—previously only affecting the rich, who could afford to consume a meat-oriented diet—has become the main health problem for developed countries.

The most heavily promoted and popular products of industrial agriculture are damaging to human health. This includes animal products such as meat, fish, dairy, and eggs, but also processed or refined plant foods such as sugar, refined flours, and refined vegetable oils as well as tobacco products. The rise of these foods has led to rising health care costs in the United States, which has in turn become an explosive and contentious political issue. The United States spent 17.1 percent of national GDP on health care in 2013, much more than comparable industrialized countries—but surprisingly, health outcomes were substantially *worse* in the United States compared to these other countries.[286]

The *administration* of health care, slanted toward benefiting pharmaceutical and insurance companies, is likely a significant cause of rising health costs, and this has led to demands for serious reform of the health care system in the United States. But nutrition is just as important, or even more so. The ultimate cause of medical costs, after all, is that people are sick.

....
286 Squires and Anderson, "U.S. Health Care from a Global
    Perspective."

The major killer diseases in the United States are heart disease, strokes, colon cancer, prostate cancer, breast cancer, lung cancer, hypertension, diabetes, and obesity. Smoking and lack of exercise are responsible for some of these problems, notably lung cancer. However, the dominant factor is our diet: high in animal products such as meat, fish, and dairy, and high in refined and processed foods such as sugar and oil. They don't occur with nearly the same frequency in less industrialized countries that do not consume as many animal products and refined foods. Nor did they occur in China, Japan, and India fifty to one hundred years ago, but as those countries have adopted Western diets, they occur with increasing frequency.

Vegetarianism, veganism, and whole-food, plant-based diets

Are plant-based diets the answer? Scientific advocates of food reform generally don't talk about veganism, but about a whole-food, plant-based (WFPB) diet. For optimum health, it is necessary to eliminate or sharply reduce consumption of refined foods as well as animal products.

Vegetarianism as a practice has existed since ancient times, though the specific term *vegetarian* was only coined in the mid-nineteenth century. It came into general use in 1847 after the founding of the first vegetarian organization in Britain. It is defined as the practice of not eating meat, fish, or fowl in any form, with or without the addition of dairy products and eggs.

Veganism is a subset of vegetarianism. It excludes dairy products, eggs, and honey from the diet as well. It further excludes any exploitation of animals (as in leather, wool, silk, or animal experimentation). The Vegan Society (based in the UK) originated the use of the term *vegan* in 1944, with the founding of the society. Today, it defines *veganism* as follows: "Veganism is a way of living that seeks to exclude, as far as is possible and practicable, all forms of exploitation of, and cruelty to, animals for food, clothing, or any other purpose."

The practice of vegetarianism has appeared spontaneously and independently in different cultures, in ancient Greece and ancient Israel as well as in the eastern religions of Hinduism, Buddhism, and Jainism. Ancient advocates include Pythagoras, Plutarch, and Porphyry, and lists of modern advocates typically include Gandhi and Tolstoy. The practice of veganism is harder to trace historically because of the uncertainty attached to the definition. In ancient times many people were on a spare vegan diet most of the time but relied on animals for transport and agricultural labor. Whether there were true vegans in ancient times revolves around whether these practices necessarily involve "exploitation" and "cruelty." Arguably, the modern concept of veganism corresponds to the first precept of Buddhism: not to cause suffering to any sentient creature.

The phrase "plant-based diet," in many circles (especially in the United States) is just another name for a vegan diet; the diet is vegan, but people adopting it may or may not abstain from exploitation of animals in other ways. But there is no official definition and some people use the phrase in a looser way; some consider "plant-based diets" to be *based* on plants, but not necessarily *restricted* to plants (Brown et al., "What Does 'Plant-based' Actually Mean?"). It might, for example, feature use of meat or fish once a month or once a week, as long as animal products aren't an essential part of the diet. As used in *this* book, however, a "plant-based diet" is the same as a "vegan diet."

The whole food, plant-based diet excludes the use of refined products such as sugar and vegetable oils. But like the term "plant-based diet," it has no official definition. There are different opinions as to what counts as a "refined food" and how much tolerance there is for occasional deviations from the plan. Most people would say whole wheat flour and rolled oats aren't refined foods, but white flour and cane sugar are. Typically, there is also more tolerance for occasional deviations related to refined foods (eating cookies with white flour and sugar once a week) than there is for occasional deviations related to animal foods (eating cheese once a week), but the usage depends on the context.

The concepts of vegetarianism and veganism have evolved and are still evolving. They originated as *ethical* concepts, and as such didn't address refined plant foods or health matters generally. Modern advocates of nutritional reform typically talk about a whole-food, plant-based diet, which is a *nutritional* concept. It is therefore *less* strict than veganism in that it might allow exploitation of animals for non-dietary purposes, but *stricter* than veganism in that it forbids consumption of refined plant foods.

Epidemiological data

Epidemiology (population studies) provided much of the impetus for the connection between diet and disease. But, as we will see below, the clinical evidence for a dietary connection has now accumulated to the point where it is overwhelming and conclusive. Many of these diseases overlap with each other both in their origins and manifestation. Heart disease, stroke, and erectile dysfunction are all associated with the same underlying disorder: atherosclerosis. Atherosclerosis, in turn, is connected with hypertension, diabetes, and obesity. There is now an abundant literature on degenerative diseases, and interested readers should check out the evidence for themselves.[287]

Wartime food restrictions in the First World War and Second World War gave us important clues that disease and diet were connected. Overall death rates, heart disease, and diabetes

---

287 For example, T. C. Campbell and T. M. Campbell, *The China Study*; Davis and Melina, *Becoming Vegan*; Greger and Stone, *How Not to Die*; Esselstyn, *Prevent and Reverse Heart Disease*.

dramatically dropped in Denmark, Norway, Austria, England, and Wales as a result of the more plant-based diet adopted out of the necessity of wartime restrictions.[288] After the Second World War, pioneering physiologist and diet researcher Ancel Keys noticed this effect of wartime restrictions and did further research. His "Seven Countries" study showed a link between atherosclerosis, saturated fat, and blood cholesterol levels. It also showed a negative association of heart disease with consumption of fruits, vegetables, bread, pasta, and olive oil. Keys wasn't a vegetarian, but he advocated for a low-fat diet and plenty of exercise. He practiced what he preached and lived to be one hundred years old.[289]

Other studies, including the Framingham Study, have amplified and supported the general direction of Keys's research.[290] Probably the most famous such epidemiological study was the China–Cornell–Oxford Project, popularly known as "The China Study." It was based on an extensive study of dietary habits and disease in the 1980s in China and resulted in a popular book titled *The China Study* co-authored by T. Colin Campbell, one of the coordinators of the project.

The strength and originality of the China Study lay in the wide variety of diets eaten in China—with some subjects eating mostly or entirely traditional Chinese foods like rice, while others followed a more Western diet high in fat and animal products. There was wide variation not only in the county averages of fat and fiber intake (which from lowest to highest varied sixfold and fivefold, respectively), but also in measured values of blood cholesterol, blood beta-carotene, and blood lipids. In the United States at the time, by contrast, eating styles

288 Akers, *A Vegetarian Sourcebook*, p. 51, and T. C. Campbell and T. M. Campbell, *The China Study*, p. 151.
289 Andrade et al., "Ancel Keys and the lipid hypothesis."
290 T. C. Campbell and T. M. Campbell, *The China Study*, pp. 114–115.

were much more homogenous.

The China Study found a wide variation in the occurrence of different diseases, broadly distinguished between diseases of poverty and diseases of affluence. Diseases of poverty (pneumonia, tuberculosis, parasitic disease) were due to inadequate nutrition or bad sanitation. Diseases of affluence include the big killers in Western countries: heart attacks, strokes, diabetes, lung cancer, colon cancer, breast cancer, and leukemia. The primary culprits here were foods of animal origin.

Clinical tests and the reversal of diseases of affluence

What is the precise mechanism in causing diseases of affluence? It is important to bear in mind that the evidence connecting a Western diet—and animal products specifically—to most of these diseases of affluence is stronger than our knowledge of the exact mechanism.

Experimental evidence shows that diet can not only halt the progress of these diseases but *reverse* them. The Kempner Rice Diet, initiated by Dr. William Kempner at the University of North Carolina in 1939, consisted of white rice, fruit, juice, sugar, and vitamin pills. Kempner's methods were questionable and the foods were nutritionally deficient (white rice, juice, and sugar?)—probably the reason vitamin pills were added. But amazingly, this diet *did* get positive results. Kempner was able to reverse or reduce heart disease, kidney failure, hypertension, and other problems. Later medical advocates of the whole-food, plant-based diet essentially made the Rice Diet healthier: they added vegetables and whole-grain foods and got rid of the refined sugar and white rice. But Kempner's studies showed that animal products had a decisive influence on these diseases *independent* of the effects of refined foods.

Dr. Caldwell Esselstyn and Dr. Dean Ornish were pioneering advocates of the use of a WFPB diet not only to help prevent cardiovascular disease, but also to reverse disease already present. Esselstyn and Ornish based their similar programs on peer-reviewed scientific research. They started with patients who already had advanced heart disease. Many had suffered heart attacks or undergone such standard surgeries as angioplasty or heart bypass surgery. Their subjects ate an almost entirely vegan and WFPB diet. (Ornish allowed a very small amount of animal foods but added exercise and meditation.)

The results of both programs were striking and positive. Within the first year, Ornish's patients experienced substantially fewer attacks of angina; after five years, 99 percent had experienced halt or reversal of their artery disease. Esselstyn had similar excellent results. Pain from angina was universally less (or disappeared completely). After twelve years, seventeen of his eighteen patients had experienced no coronary events after beginning the program; one patient didn't follow the program and required surgery but survived. Angiograms showed dramatic evidence of reversal of the disease and clearing of the arteries.[291]

Diabetes can also be effectively treated and reversed with a WFPB diet. James Anderson put twenty-five patients with type 2 diabetes on a largely WFPB vegan diet and found that within a matter of weeks, twenty-four out of the twenty-five didn't need any insulin medication.[292] In 2007, Neal Barnard and his associates did a trial in which people suffering from type 2 diabetes were randomly assigned to consume either the American Diabetes Association's standard diet or a low-fat, low-sugar vegan diet. Both groups showed improvement, but the vegan group was able to lower medication more, lost more

291 Esselstyn, *Prevent and Reverse Heart Disease*, pp. 89–92.
292 T. C. Campbell and T. M. Campbell, *The China Study*, p. 152.

weight, and lowered LDL cholesterol levels more.[293]

Rates of many common cancers in populations on Western diets are often many times higher than rates in populations on a mostly plant-based diet. It is hard or impossible to reverse many cancers with diet alone after diagnosis, but a WFPB diet helps *prevent* cancer, and sometimes diet can help in the treatment.[294]

The same general observations made about heart disease, diabetes, and cancer can be repeated for many of the most common diseases in the industrialized countries: obesity, kidney diseases, osteoporosis, multiple sclerosis, and others. Vegetarians and vegans consistently have been shown to suffer less from obesity. Multiple sclerosis (MS), although an auto-immune disease, shows the same patterns and associations seen for diseases of affluence. While MS can't be reversed through diet, its progress can be slowed or halted on a low-fat vegan diet. T. Colin Campbell, one of the directors of the China Study, suggests that universal adoption of a WFPB diet could reduce health care costs by at least 60 to 80 percent.[295]

Attempts to confuse the issue

Knowledge of the benefits of a WFPB diet are spreading. But for many in the general public, this information is a surprise.

The agricultural and pharmaceutical industries have well-funded lobbies that often seek to prevent such information from spreading. "Denialists" are as notable in nutrition science as they are in climate science. It is telling that the significant lifestyle issues that *have* received attention—smoking and lack of exercise—are those that don't endanger nearly as many of these industries' interests. The tobacco industry is less powerful than

293 Barnard et al., "A low-fat vegan diet improves glycemic control."
294 Meyer et al., "Dietary fat and prostate cancer survival"; Greger and Stone, *How Not to Die*, chapters 4, 11, and 13.
295 Our Hen House, "Interview with Dr. T. Colin Campbell."

the agricultural and pharmaceutical industries, but it still put up quite a bit of resistance for decades (and tobacco smoking still flourishes in many parts of the world). Scientific research showing the benefits of exercise, though, don't threaten any industry at all, which partially explains why today fitness centers have sprouted up all over the country.

Industry efforts to downplay health impacts seek to confuse the issue by a variety of tactics. They may seize on an ambiguity, spread misleading or outright false information, influence the publication of deliberately misleading scientific studies, or selectively quote credible articles, all to create the impression of "controversy." For example:

1. When the connections between saturated fat and heart disease first emerged in the 1960s and 1970s, some advocated consumption of low-fat milk, refined vegetable oils, and switching from beef to chicken—foods lower in saturated fat than beef, whole milk, and eggs. But it turned out that these strategies didn't work. Cholesterol levels of people switching from beef to chicken or drinking low-fat milk didn't change. Cancer rates of people switching to vegetable oils also didn't change; if anything, they seemed to increase. More recently, others have repackaged this confusion to deny that there is any connection between saturated fat and heart disease at all.[296]

This confusion arises because our knowledge of the foods that are healthy is stronger than our knowledge of which constituents are involved or the precise mechanism. Saturated fat *does* facilitate heart disease,[297] but so does animal protein; and refined oils also contribute to cancer. Saturated fat is closely associated with all these diseases because *saturated fat closely tracks consumption of animal products*. Animal products are

....

296 See, e. g., *Plant Positive*, "How Time Magazine Sacrificed Its Standards."
297 Greger and Stone, *How Not to Die*, pp. 28–29.

often responsible even though we don't always understand the exact mechanism involved.

2. In 1977, US Senator George McGovern chaired the Senate Committee on Nutrition and Human Needs, which issued the first federal nutrition guidelines for Americans in January of that year. The McGovern committee looked at the evidence and issued a draft that contained some common-sense recommendations, including to eat less meat, salt, and sugar. The response from the food industry was overwhelming and negative. According to Dr. Mark Hegsted, a prominent Harvard nutrition researcher, "The meat, milk and egg producers were very upset." The meat industry "reacted rather violently" to the report, complaining about the report's constant association of meat consumption with degenerative diseases, and insisting that promoting animal products was the "honorable and morally correct diet course." The National Dairy Council demanded that future reports first get the approval of the food industry. Not only were the McGovern committee's recommendations reversed in the final report, but the committee itself was disbanded and made part of the Senate Committee on Agriculture, and several members of the committee were defeated for re-election.[298]

3. In April 1996, Oprah Winfrey had former rancher Howard Lyman as a guest on her show to talk about mad cow disease. After Lyman's comments warning of the dangers of feeding cows back to cows, Oprah said, "It has just stopped me cold from eating another burger!" Cattle prices dropped and cattlemen blamed the *The Oprah Winfrey Show*, suing Winfrey and Lyman for damages. Winfrey and Lyman eventually won the case, but they were tied up in the courts for five years for doing nothing more than giving their opinion (in fact, a very well-justified opinion).

....

298 Greger, "The McGovern Report."

4. People on the Dietary Guidelines Advisory Committee for the "Dietary Guidelines for Americans" frequently have ties to the food and drug industries. On the 2000 advisory committee, for example, "members had past or present ties to: two meat associations; four dairy associations and five dairy companies; one egg association; one sugar association; one grain association; five other food companies; six other industry-sponsored associations; two pharmaceutical associations; and 28 pharmaceutical companies."[299]

COVID-19 and emerging infectious diseases

While chronic diseases due to eating animals have a huge impact and have gathered most of the attention so far, emerging infectious diseases can also have a devastating effect, as demonstrated by the COVID-19 virus pandemic beginning in 2020.

"Emerging infectious diseases" are different from the "old" infectious diseases such as smallpox, polio, diphtheria, whooping cough, cholera, measles, and mumps. These latter diseases have been brought under control or even eradicated through antibiotics, improved sanitation, immunization, and other new medical approaches, creating optimism that infectious diseases might be largely eliminated. But emerging infectious diseases have greatly dented this optimism. The WHO *World Health Report 1996* worried that despite sizable progress against such diseases as polio, leprosy, and river blindness, that "we also stand on the brink of a global crisis in infectious diseases," and that "fatal complacency ... is now costing millions of lives."[300] More than ten years later, the *World Health Report 2007* stated that "Infectious diseases are now spreading geographically much

299 Herman, "Saving U.S. Dietary Advice from Conflicts of Interest."
300 World Health Organization. *The World Health Report 1996*, p. v.

faster than at any time in history ... [and] appear to be emerging more quickly than ever before."[301] "The concept of 'emerging infectious diseases' has changed from a mere curiosity in the field of medicine to an entire discipline," concluded Dr. Michael Greger.[302]

COVID-19 is the most recent and spectacular example of these emerging infectious diseases. These infectious diseases are "new" in the sense that they were previously entirely unknown or infrequent in human populations. They are increasingly frequent today because of (1) increasing human intrusion into animal habitats and (2) the practice of crowding animals together on factory farms and in "wet markets" (live animal markets). Neither was widely practiced before about 1900.

COVID-19 likely originated in bats and was spread from a wet market in China.[303] What is especially problematic about wet markets is that not only are animals crowded together, but animals of *different species* are crowded together, allowing opportunities not only for mutation but also for interspecies transmission.[304] We don't know the precise chain of transmission of COVID-19 through various animal species and finally to humans, but pangolins seem to be the immediate source of the virus.[305]

It is important to understand that COVID-19 isn't an anomaly. It's part of a general pattern. Almost all these emerging infectious diseases have their origins with animals; most of them are viruses.[306] Other diseases include:

....

301 World Health Organization. *The World Health Report 2007*, p. x.
302 Greger, *Bird Flu*, p. 86.
303 Huang Chaolin et al.,"Clinical features of patients infected with 2019 novel coronavirus."
304 Chan, J. F. et al., "Interspecies transmission and emergence of novel viruses."
305 Zhang et al., "Probable Pangolin Origin of SARS-CoV-2."
306 Greger, *Bird Flu*, pp. 86–87.

1. AIDS (acquired immune deficiency syndrome). With antiretroviral therapy, AIDS/HIV is killing fewer people, but it has still killed 36 million people since the beginning of the epidemic in 1981.[307] The CDC states that this was an animal disease transmitted to humans when humans ate butchered chimpanzee meat.[308]

2. Avian influenza or "bird flu." Influenza is a virus that originated in poultry but has mutated to infect humans. It consists of a variety of different types. Some only affect poultry or only affect cattle, but some affect humans. Each type in turn mutates from year to year, thus each year's seasonal flu is always slightly different from that of previous years.

These mutations can lead to a wide variety of different viruses. Given enough chances, a virus can mutate to be able to cross the species barrier, so that a virus that previously only affected poultry can spread to humans or become more deadly. With substantial increases in the number of factory-farmed chickens in the past few decades, we are giving the virus abundant opportunities to do exactly this. It is an accident waiting to happen.

Perhaps the most notorious example of such a disease was the pandemic of 1918. This disease was probably an avian flu originating in domesticated poultry in Kansas. It then migrated to Europe with American troops fighting in World War I and spread all over the world, killing fifty million people.[309] Other, much less serious flu pandemics happened in 1957 (the "Asian flu"), 1968 (the "Hong Kong flu"), and 2009 (the "swine flu"). In 1997, a particularly deadly form of flu spread directly from chickens to humans, but fortunately, it didn't transmit very

307 Cichocki, "How Many People Have Died of HIV?" This article is periodically updated and was accessed on June 9, 2022.
308 Centers for Disease Control and Prevention, "About HIV/ AIDS."
309 Jordan, "The Deadliest Flu."

well from human to human. Had a small additional mutation occurred that would have made this possible, the 1918 pandemic could easily have been eclipsed;[310] mortality from the 1997 virus was 33 percent[311] as opposed to the 1918 mortality of perhaps as high as 10 percent.[312]

3. *Ebola hemorrhagic fever* is a highly lethal but fortunately rare disease. Similar to AIDS, it likely came from great apes that were hunted for food.[313]

4. *Severe Acute Respiratory Syndrome (SARS)* is a coronavirus similar to MERS and COVID-19. It was rather deadly (just under 10 percent mortality rate) but fortunately was quickly contained; there have been no reported cases since 2004.[314] It originally came from bats, but the immediate transmission was associated with small carnivores such as civets.[315] Like COVID-19, its spread is linked to wet markets.[316]

5. *Middle East Respiratory Syndrome (MERS)* is a coronavirus similar to SARS and COVID-19. It originated on the Arabian Peninsula, possibly from camels, though the original virus came from bats (like SARS and COVID-19). It is highly deadly, killing about 30 to 40 percent of the people it infects, but fortunately hasn't spread very far.[317]

....

310 Greger, *Bird Flu*, pp. 35–39.

311 Greger, *Bird Flu*, p. 37.

312 Taubenberger and Morens, "1918 Influenza," states that about 50 million people died out of 500 million infected, but because of uncertainty gives the case fatality rate as >2.5 percent.

313 Karesh et al., "Wildlife Trade and Global Disease Emergence."

314 Centers for Disease Control and Prevention. "Severe Acute Respiratory Syndrome."

315 Karesh et al., "Wildlife Trade and Global Disease Emergence."

316 Webster, R. G., "Wet markets."

317 Centers for Disease Control and Prevention, "Information about Middle East Respiratory Syndrome (MERS)."

6. *Antibiotic resistant bacteria.* Antibiotics in the United States (as well as in much of the rest of the world) have been routinely fed to livestock on factory farms to promote animals' growth for decades. This has led to antibiotic resistance among the bacteria infecting these animals, and these bacteria can then infect humans. Researchers estimated that about 80 percent of all antibiotic use in the United States is for food animals. Overseas, Brazil, China, Russia, India, and South Africa are all expected to greatly expand their use of antibiotics in food animals.[318]

The CDC estimates about half a million cases of antibiotic-resistant gonorrhea in 2019, more than twice as many as in 2013. There were more than 400,000 cases of drug-resistant campylobacter infections, more than 200,000 of drug-resistant salmonella infections, and a total (in case anyone is keeping count) of more than 2.8 million drug-resistant infections.[319] Most of these diseases were easily treatable with antibiotics before their widespread use in livestock.

Conclusions

A smaller, more limited whole-food, plant-based (WFPB) agriculture, taking much less agricultural land and consuming far fewer natural resources, would benefit human health as well as the health of the ecosystem. By reducing our contact with wild animals and opportunities afforded by animal agriculture for virus mutations and antibiotic resistance, a more plant-based agriculture would reduce the likelihood and frequency of pandemics such as the 1918 flu pandemic and the 2020 COVID-19 pandemic.

We all now have direct experience of the sizable disruptions

....

318 Van Boeckel et al., "Global trends in antimicrobial use in food animals."

319 Centers for Disease Control and Prevention, *Antibiotic Resistance Threats in the United States, 2019.*

that a worldwide pandemic can occasion. We cannot risk such consequential accidents waiting to happen. By reducing the incidence of chronic degenerative diseases such as heart disease, cancer, obesity, and diabetes—much of which is caused by eating animals—we would not only improve human health, but also reduce medical expenses that are a significant burden on both consumers and the economy. T. Colin Campbell comments: "The triumph of health lies not in the individual nutrients, but in the whole foods that contain those nutrients: plant-based foods."[320]

---

320 T. C. Campbell and T. M. Campbell, *The China Study*, p. 94.

CHAPTER EIGHTEEN

HOW MANY PEOPLE CAN THE EARTH SUPPORT?

Population, the economy, and the biosphere

Population is a complex and sensitive topic. Our population levels are likely already unsustainable; we need a substantial *reduction* in the bloated human presence on the planet. Assuming humans proceed diligently on other issues, we may have quite a while to achieve this aim before Mother Nature does it for us—but we can't dawdle.

We face *multiple* limits to growth, not just one. The optimum human population should really be close to the *last* major parameter of an ecological civilization to be determined because it rests on all the other parameters combined (agricultural land, energy use, resources, biodiversity, social structure, and social preferences). In this chapter we just want to get a better idea what this optimum human population for our planet might be going forward, postponing for the moment how we could get there.

Any estimate of maximum sustainable population *also* presupposes a certain social structure. Resources probably aren't going to be divided equally among all the nation's (or the world's) citizens. We *may* also need to include a generous surplus of resources to be dedicated to maintenance of the elites. All advanced societies, so far, have had an elite class that had substantial material privileges over the common worker, even the former Soviet Union. This adds another layer of complication and shows how slippery the concept of "sustainable population" is. On top of that, we probably do not want to push human numbers to the absolute limit, to avoid the consequences of a miscalculation. The *optimum* sustainable human population is likely to be less than the *theoretical maximum* sustainable population.

Optimum human population

Experts offer widely differing estimates of the optimum or maximum sustainable human population. One list of more than sixty population estimates, made between 1679 and 1994, gave estimates mostly ranging from about 0.5 billion to 14 billion but also some fantastical estimates of 50 billion, 300 billion, 800 billion, or 1000 billion (one trillion).[321] Most such estimates appear to bypass soil erosion, energy supplies, or biodiversity concerns—exactly the parameters we're most interested in.[322]

Even progressive environmentalists who seem aware of these sorts of limitations have put forward estimates that vary considerably:

321 Cohen, Joel E., *How Many People Can the Earth Support?*, Appendix 3, pp. 402–418.
322 In Joel Cohen's book *How Many People Can the Earth Support?*, there appears to be only the briefest mention of soil erosion, energy supplies, and biodiversity concerns. These would surely would have been discussed if they bore on any of the numerous estimates that Cohen cites in his book.

1. Dr. Jack Alpert, director of his Stanford Knowledge Integration Laboratory, suggests less than 100 million.[323]

2. Paul Ehrlich, co-author of *The Population Bomb*, has given different estimates during his long public career. In 1971, he suggested the optimum number was between 500 million and 1.2 billion.[324] In 2012, he gave a slightly more optimistic number of from 1.5 to 2 billion, assuming lifestyles of "big active cities and wilderness."[325]

3. David Pimentel, noted writer on ecology and emeritus professor at Cornell, is the lead author in a paper suggesting an optimum human population of 2 billion humans would be sustainable in about a century when we have "run out" of fossil fuel energy. The paper takes into account soil, water, and energy resources and assumes a European standard of living for everyone, with sustainable use of resources and a halt to *all* current land degradation.[326]

None of these estimates can be dismissed out of hand. Do we really know anything? Fortunately, from an analytic point of view, we don't have to go through all possible parameters in detail. Our friend, "Liebig's law of the minimum" (see chapter 4) suggests that population is limited by the single *relatively* scarcest factor. What we need to do is to identify that scarcest single resource.

---

323 Stanford Knowledge Integration Laboratory, "The earth's sustainable population is below 100 million."
324 Ehrlich, "The Population Crisis," p. 8.
325 Vidal, "Cut world population and redistribute resources."
326 Pimentel, Whitecraft, Scott et al., "Will Limited Land, Water, and Energy Control Human Population Numbers in the Future?"

What is the limiting factor on population?

To really assess a population estimate, we need to know the underlying logic. Following "Liebig's law of the minimum," what is the ultimate limiting factor on human population?

There are many factors to choose from! The easiest metrics to consider would be those pertaining to energy, food, and biodiversity. Other factors may be involved as well; people have raised plausible questions about phosphorus, oil, groundwater, rare earth minerals, and many other "ingredients" needed for modern civilization. Because of the inherent complexity of the problem, it is likely easier to look for parameters that are *not* in short supply and to build a model of sustainable human population from there, rather than working backward from our current bloated economy looking for ways to cut back.

So we will start with energy, food, and biodiversity. We've already discussed them in moderate detail; and any one of them, *by itself*, could pose an upper limit on sustainable human population. Are there resources, sufficient to power civilization, that we can clearly count on, given these three metrics?

1. *Energy use.* Considering this metric assumes that the ultimate limiting factor will be *energy* supply. Fossil fuels are already off the table as a permanent energy source, as climate change is a key factor and fossil fuels are finite anyway. What energy source do we have that is renewable? Biofuels have worked for thousands of years; before the Industrial Revolution, the entire world ran on renewable energy, namely biofuels in the form of wood.[327]

While biofuels are renewable, they don't offer a lot of energy. Can we increase our supply of renewable energy with other energy sources? Hydroelectric power offers some help but is

327 This should be an obvious point, but thanks to Alice
   Friedemann, *Life After Fossil Fuels*, for pointing it out.

running at maximum output already and only fulfills a small fraction of our current energy use. Modern renewables such as solar, wind, and nuclear power are all limited in different ways and will likely offer us much less energy than we use now.

It's possible that solar, wind, and nuclear (or some combination), while technically finite, could create additional energy to support substantially higher population levels for thousands of years. (Sunlight and wind are functionally infinite, but the materials to build solar and wind technology are not.) But this has most definitely *not* been demonstrated. Until we know for a fact that these other energy sources are truly renewable and sustainable, we should assume that our safest sustainable bet is wood and biofuels and make corresponding population plans.

The last time the world economy ran on biofuels, the population was about 650 million, less than 10 percent of current levels. If we decide this is our "safe" optimum population, we have some time to get to this objective, and with hydroelectric power and some modest use of solar, wind, and nuclear, we may be able to improve living conditions and increase population levels further. But pre-industrial population levels seem to be the "safe" benchmark here in terms of energy supply.

2. *Soil erosion.* Or we could assume that *food* is the ultimate limiting factor and that soil erosion limits food. If we assume that soil erosion worldwide is no more than the relatively low "American" level—which is still *at least* ten times that of the natural process of soil formation—then it would seem that we will have to leave *at least* 90 percent of all farmland fallow each year to allow for soil formation sufficient to compensate for soil erosion. Even if the entire world goes vegan (a vastly more efficient use of land), how many people could be supported on 10 percent of the world's cropland?

One pessimistic (but plausible) approach is offered by Dr. Jack Alpert.[328] He argues that for social justice, we must raise everyone's per-capita consumption to the current level of the top 20 percent of the population. Doing that, however, would quadruple the total resource usage, so we can sustainably support only one-fourth of the population. On top of that, soil erosion means we must leave fifteen acres fallow for every acre we cultivate; thus we can sustainably support only one-sixteenth of the population for a given amount of agricultural land. Combining these two requirements means that we can support only one-sixty-fourth of the population in a world combining social justice with soil sustainability. When Alpert did these calculations (in 2003), he assumed that world population was 6.4 billion, so he concludes that a sustainable and socially just society could only have a maximum population of 100 million, about the world population at the time of the Buddha and Pythagoras. In this view, the world (at 600–700 million) was already overpopulated before the Industrial Revolution began!

We could walk back some of Alpert's population pessimism as follows. For social justice *in terms of food*, we don't want to raise resource consumption to the destructive level of the top 20 percent of the world. Rather, we would want the reverse: for their own health, the elites should be reduced to the food-consumption levels of everyone else. Universal adoption of plant-based diets, even by elites, is quite plausible. Everyone's basic nutritional needs are about the same, and we should get rid of most livestock agriculture for environmental and health reasons anyway.

But we still have the problem of soil erosion. If we follow Alpert's suggestion to leave fifteen acres fallow for every acre cultivated, that would leave us with a ballpark estimate of a supportable population of 500 million people (8 billion/16). We

328 Alpert, "The earth's sustainable population is below 100 million."

have some wiggle room, but that puts us near population levels just before the Industrial Revolution.

We may be able to add a few optimistic but not implausible assumptions to boost this estimate even further. If we assume that we will now have all the knowledge gained in the twentieth and twenty-first centuries and that books and education, at least for the scientific elite, will continue to exist, then we can make many people into organic farmers and use the latest available sustainable farming techniques. It might also turn out that most soil erosion occurs on a small portion of highly erodible agricultural land that is being farmed by the poor only out of desperation. By simply taking this land out of production, we could further decrease soil erosion. If we reduce soil erosion to the point that we only need to leave three to five acres fallow for every acre cultivated, all this combined might boost sustainable population closer to Ehrlich's and Pimentel's estimates of 1.5 to 2 billion; and in fairness to Pimentel, he puts special emphasis on halting land degradation.

3. *Megafauna biomass.* Finally, we could assume that *biodiversity* is the limiting factor. For 99,500 of the past 100,000 years, megafauna biomass has been at, or even below, 200 billion kilograms (see chapter 5). It is only in the past 300 to 500 years, since the Industrial Revolution, that megafauna biomass has drastically increased more than sevenfold—a biologically unprecedented event in recent Earth history. Most of this increase is attributable to livestock, most of the rest to humans; only a small sliver of the total (less than five percent) belongs to wild animals like elephants and antelope. It's the humans and livestock who are the biodiversity problems in this situation.

Because this 200-billion-kilogram level was stable for so long, we could assume that this is the "natural" biological limit of megafauna biomass—that is, the limit of megafauna biomass in the absence of humans. We could argue that this "natural" limit is also the *sustainable* limit. This argument seeks to cut through all the complex information about soil, water, and energy, to find a *single* metric that best approximates what we are looking for. The earth has limited land area and, given the limitations of soil and sunlight, only a finite amount of plant matter can be grown. All megafauna require plants for food, either directly or indirectly (the carnivores will need to eat plant-eating animals). Therefore, this argument goes, a *sustainable* agriculture would support megafauna biomass of humans and wild animals that approximately equals this "natural" limit of megafauna biomass. Trying to extend this biomass inevitably will strain our ecosystem somewhere: soil erosion, mineral depletion, biodiversity collapse, energy supplies, or something else.

If we were to revert somehow to 200 billion kilograms of megafauna biomass and generously dedicate 100 billion of this total to wild megafauna (in accordance with the "Half-Earth" principle), this gives us about 1.67 billion plant-eating humans and a wide area for biodiversity and wilderness to flourish.

This crude ballpark estimate makes a number of assumptions about biodiversity. What we really want is biodiversity of *everything*, not just megafauna, which by biomass are a fairly small proportion of all animals. We are assuming that if we look out over the landscape and see a broad diversity of megafauna, similar to the landscape of 100,000 years ago, that this is a good first approximation of the rest of the environment. If by contrast humans, or humans and livestock combined, are practically the only megafauna around, then it is likely that we have a biodiversity problem and that some species, smaller creatures as well as megafauna, are being eliminated to make way for humans.

None of these estimates give us any degree of precision about the answer we wanted—an estimate of the optimum human population. They are all back-of-envelope population estimates, and all of them can be debated and challenged. You, the reader, can probably think of ways to do this yourself. Scientific consensus on this question is extraordinarily unlikely; there are just too many unknowns.

But what seems indisputable is that *all* the resources currently needed for our civilization are degrading, and not to a trivial degree. Even soil, water, and plant matter are in serious decline, and we haven't even gotten to supplies of exotic things like lithium, neodymium, and industrial heat.

In the meantime, we are playing with fire. One thing that seems close to being "settled science" is this: habitat destruction leads to species extinction. Today, we are doing more than systematically destroying wildlife habitat with soil erosion, biodiversity destruction, chemical pollution, and energy depletion; we are also systematically degrading our *own* habitat. We should be thinking about avoiding our *own* extinction.

Is there a minimum population needed for civilization?

There is an additional "imponderable" difficulty: the possibility of human population falling too *low*. Is there a minimum number of humans needed to sustain any kind of reasonably advanced civilization, capable of preserving most of our current realm of scientific knowledge?

In the science-fiction novel *Station Eleven*,[329] the characters face this sort of challenge when a deadly plague wipes out almost all human beings. The survivors don't know how to make many of the common devices they previously took for granted—like computers, light bulbs, and telephones—even with libraries and books still around and much less competition for natural resources (with most other humans dead). But if you were a survivor in the world of *Station Eleven*, would you be able to

329 By Emily St. John Mandel (2014).

figure out how to make a working light bulb by reading books at local or university libraries?

The minimum number required to preserve the standard of living available in the seventeenth century was quite small. European colonists to the New World could create a more or less self-sustaining colony with just a few hundred people, mostly farmers, though at first they depended on regular inputs of information and supplies from the Old World. The world of *Station Eleven* seems to have returned to this seventeenth-century level of technological expertise.

Information is the critical resource here. In our current world, experts can advise us how to perform specialized activities. The production of washing machines and other mundane economic activities require an entire support structure that includes many specialties, such as mining, machine assembly, and electricity.[330] It is doubtful that any single person knows *personally* about all the various aspects of modern clothes-washing from beginning to end.

Modern information networks can replace, to a certain extent, the need for knowledgeable experts. But information technology leaves us vulnerable in new ways. It is dependent on scarce rare earth minerals and increases societal complexity. It leaves society vulnerable to various abuses that limit its usefulness, such as spam, computer viruses, "fake news," cyber-bullying, social media, politics, war, and fraud.

There is also the danger of a data disaster such as a gigantic solar storm. In 1859 there was such a storm, called the "Carrington Event" after British astronomer Richard Carrington who noted solar flares at the same time telegraph equipment was disrupted. This solar storm had only a small effect on the relatively nontechnical world of 1859, but a similar storm might devastate our computer equipment in the twenty-first century.

---

330 Humes, *Door to Door* discusses coffee and aluminum cans.

Unguarded electronic equipment might be destroyed outright; recovery from such a disaster might take up to ten years and trillions of dollars even in our modern world.[331] The devastation of such a storm, in a technologically advanced society, might dwarf the impact of the burning of the library of Alexandria in ancient times, especially if we turn to storing all our records digitally. Relying on information technology to reduce the population needed to sustain civilization, therefore, exposes us to new hazards.

Conclusions

We need to *substantially* reduce human population. The IPAT equation (see chapter 5) suggests that we can accommodate more humans by finding additional natural resources, by devising more ingenious technologies, or by reducing consumption. In these ways we might be able to accommodate more humans on a temporary basis. But there are some basic resources that, in the long run, are already stretched far beyond any sustainable level and which are indispensable to human civilization. Three of these are energy, soil, and biodiversity.

In the meantime, we are pressing hard against even more mundane resources and pressing against each other. Anyone who doubts this should perhaps think about it further while sitting in a traffic jam while breathing polluted air, or should visit one of our national parks and reflect on how little of this natural world still exists and how many people want to crowd in to see it.

....

331 Letzter, "A massive solar storm could wipe out almost all of our modern technology."

CHAPTER NINETEEN

FACING THE ECONOMY
The economics of sustainability

What economic policies will we need to ensure an appropriate scale for the economy? We will need much more than economic policies, but addressing how big the economy should or can get is in many ways the central problem of this book.

The economic context

In economic terms, "embracing limits" means setting economic policies to limit the *scale* of the economy: limiting how big the economy can get before it consumes everything on the planet, including us.

Letting the free market sort things out, while it helps in other ways, has clearly failed at limiting scale (see chapter 11). In an "empty world" with few humans and many natural resources, there are few practical limits to growth and the free market can generally operate without destroying the environment (though society might face social justice problems). But in a "full world" with scarce resources, at some point a growing economy will impinge on the very resources on which the economy (and indeed all life) depends. We can plainly see that climate change is out of control and that the Colorado River runs dry before it reaches the ocean; but businesses still compete for more oil and for scarce water. Most mainstream economists, who should be the experts on limits to growth, seem to have little awareness of the significance of these issues.

Fortunately, we aren't starting from scratch in trying to envision what such policies would be. In 2009, Herman Daly—the best known of the ecological economists—offered a set of ten policy suggestions (see Box 19-1, "Herman Daly's ten points summarized").[332] The easiest way to give an overview of the economic context of limits to growth is to summarize these ten points. Many of the proposals that Daly discusses have already been discussed by others, and some are quite well known, such as "cap and trade." Economic policies to address resource limits will need to address these issues—if not in the way that Daly suggests, then in some similar way that achieves a similar effect.

Most of these points will be discussed in this chapter. (Daly's point 9 will be dealt with in chapter 20, "Facing future generations," and points 3 and 4 will be dealt with in chapter 21, "Resistance, work, and technology.") Some of his suggestions are straightforward, like taxing carbon emissions; others are more esoteric, like creating a "multilateral payments clearing union." The important thing to remember is that they all aim to do either or both of two things: (1) limit or regulate the physical size of the economy, and/or (2) reduce global economic inequality.

Stronger community control

To protect natural resources, society needs to protect them *directly*. This implies stronger governmental intervention to protect resources, whether these resources are wilderness, oil, or the atmosphere.

....
332 H. Daly, "From a Failed Growth Economy to a Steady-State Economy."

Protecting natural resources doesn't necessarily imply *nationalization* of all natural resources, but it *would* require that the right to exploit them would be dependent on the government in some form. This would imply control over fossil fuel and mineral resources, agricultural land, wilderness areas, forest land, water rights, and rights to emit greenhouse gases.

Cap–auction–trade (item 1 in Box 19-1) is one of the best known of all ecological economics proposals. To protect the atmosphere from greenhouse gas emissions, we could set an annual upper limit on emissions and auction off permits to emit them (e.g., by burning coal) up to that limit. The permits would be auctioned off then could be traded once purchased, so that the company needing to burn coal the most (or, at least, willing to pay the highest price) would wind up with the permit. Restrictions would be based on depletion or pollution limits— whichever threat was greater. It seems likely that depletion is a bigger threat in the case of oil, but pollution (in the form of greenhouse gas emissions) is the greater threat for coal. If we wanted to phase out fossil fuels entirely by, say, 2050, we could sell fewer and fewer permits as the date drew closer. Eventually, the permitted amount of greenhouse gas emissions would sink to zero or nearly zero.

While "cap and trade" is best known in the context of greenhouse gas emissions, this same policy could function for *all* scarce resources. For example, we could set an upper limit on water use in the Colorado River basin, figuring in the need to support wildlife in the basin as well as to dole it out for personal consumption, agriculture, and industry. We could set a limit on oil drilling or coal production based on scarcity (if scarcity is a greater concern than climate change). Similar limits could be set on farmland, grassland, forest land, water, and mineral resources. Restrictions on methane emissions should include methane from livestock as well as from the oil and gas industry.

1. Cap–auction–trade	Put caps on use of basic resources according to source (supply) or sink (pollution) constraint, whichever is more stringent.
2. Ecological tax reform	Shift tax base from labor and capital to the extraction of natural resources and pollution.
3. Limit inequality	Limit the range of inequality in income distribution—a minimum income and a maximum income.
4. More flexible work schedule	Free up the length of the working day, week, and year. Allow greater option for part-time or personal work.
5. Re-regulate international commerce	Move away from free trade and globalization by protecting economies that adopt resource-efficient and socially just policies from standards-lowering competition.
6. A multilateral payments clearing union	Downgrade the International Monetary Fund, World Bank, and World Trade Organization. Move in the direction of avoiding large capital transfers and foreign debts.
7. Move away from fractional-reserve banking	Move away from fractional-reserve banking toward a system of 100 percent reserve requirements.
8. Put natural resources in public trusts	Enclose the remaining commons of natural resources in public trusts. Price it by a cap–auction–trade system or by taxes. But deprivatize knowledge and information.
9. Stabilize population	Work toward a balance in which births plus in-migrants equals deaths plus out-migrants.
10. Reform national accounts	Separate GDP into a costs account and a benefits account. More GDP isn't always a good thing.

Box 19-1. Herman Daly's ten points summarized

Source: H. Daly, "From a Failed Growth Economy to a Steady-State Economy."

Ecological tax reform (item 2 in Box 19-1) could work in much the same way, following the slogan "tax bads, not goods." Instead of taxing income, for example, we could tax mining coal or burning gasoline—thus the most famous proposal to accomplish this, the carbon tax. This system could be set up to be revenue-neutral, by proportionally reducing taxes on things we don't want to tax, like income. A carbon tax could work in synchronization with a cap-and-trade policy on greenhouse gas emissions. We should also undo the subsidies for livestock agriculture that encourage the production of meat at the expense of land and wildlife, as well as impose a meat tax to restrict or eliminate meat consumption.

Putting natural resources in public trusts (item 8) reframes the previous two points in a different context. There are some natural resources—for example, the atmosphere—that are unowned and unregulated. Not only should privately owned natural resources (such as oil) be regulated, but open-access regimes (such as the atmosphere or fishing rights) should be regulated as well. *All* natural resources should be protected in some way. In the case of climate change and the oceans, this would need to be accomplished through international treaties, as neither the climate nor most of the oceans observe political borders.

Just to prove that ecological economics isn't all about regulation, there are some areas that Daly proposes to *deregulate*, such as intellectual property goods. Much valuable information and technology are currently privately owned due to patents and copyrights. Because information and technology (unlike physical resources such as oil and neodymium) are not depleted if more people use them, they need to be deprivatized and made more widely accessible. There needs to be some way of compensating people who come up with new technologies or new literary creations, for example, to encourage innovation; musicians and writers come to mind, as well as scientists involved in research on new cures for diseases. But the current system goes way beyond compensation and gives too much control to a few corporations.

Change the national accounting system

There's another problem that affects the scale of the economy: how we measure the size of the economy in the first place.

Measuring the health of the economy in terms of total economic activity, say by gross domestic product (GDP) or gross national product (GNP), is inaccurate for a number of obvious reasons. Some things that increase GDP are bad, while some things that do nothing or very little for the GDP are good. If someone eats meat and junk food, smokes for years, has a heart attack, and then goes in for triple-bypass surgery, all of this is positive in terms of GDP. It employs farmers, food-processing companies, pharmaceutical representatives, and surgeons. If someone adopts a plant-based diet, the only improvement to GDP may be an uptick in sales of fresh fruits and vegetables. Humans would clearly be better off, yet in GDP terms the economy would suffer.

Daly suggests that we separate GDP into a *cost* account and a *benefits* account. Then, when the economy grows but the cost side starts to equal or exceed the benefits, we need to stop economic growth because it's not worth it. *Benefits* are declining even though jobs are being created.

Economists debated the use of GDP (or GNP) as a measure of the economy in the 1970s and decided that GDP was a reasonably good proxy for human welfare. But when Herman Daly and John Cobb revisited this issue, they found that the evidence that GNP has actually increased welfare since 1980 is "probably non-existent."[333] Later, a number of policy analysts developed an alternative index known as the Genuine Progress

....

333 H. Daly and Farley, *Ecological Economics*, pp. 233-234.

Indicator (GPI), intended to measure the welfare of the population rather than the total amount of economic activity.[334] Between 1968 and 2002, the GPI and GDP of the United States progressively diverged; the GDP doubled, while the GPI stayed about the same.[335] This suggests that GDP is increasingly inaccurate at reflecting human welfare. The GPI has now been adapted in slightly different ways by a number of different people and groups. In all likelihood, we have long passed the point where costs start to exceed benefits, and we should stop trying to further grow the economy right now.

Banking and trade

The ideology of the growth economy has been thoroughly integrated into our banking and trade systems. These systems already introduce biases toward encouraging exploitation of both humans and the environment. Three of Herman Daly's points relate to this issue (points 5, 6, and 7 in Box 19-1, "Herman Daly's ten points summarized").

1. *Move away from fractional-reserve banking* (point 7). It isn't well understood by the general public, but private banks create money—not by printing it (that would be counterfeiting), but by loaning out money they don't have, which is not only legal but officially encouraged. The concept of fractional-reserve banking is that the bank doesn't have to keep enough cash on hand to meet the potential demands of all its depositors. It only has to keep a fraction of potential demands against the bank on hand in the form of cash (known as "reserve requirements").

334 The original developers of the GPI were Clifford Cobb, Ted Halstead, and Jonathan Rowe. Anielski, "The Genuine Progress Indicator."
335 L. Daly and Posner, "Beyond GDP."

Since the Second World War, this practice has worked reasonably well. Banks have managed to do this for three reasons: (1) most people repay their loans, (2) the economy has grown, and (3) depositors don't all ask for their money back at the same time. On a typical day, deposits will approximately equal withdrawals—here a check is written, there a paycheck is deposited.

There are two problems with this system: when it works, and when it doesn't. A spectacular example of the system *not* working is the epic mortgage fraud that precipitated the financial crisis of 2008. When people couldn't pay their mortgages, the banks that held them started to fail, and the whole system would have collapsed without government intervention. The FDIC can cover if one or two banks go belly-up, but they obviously can't help if there is a run on *all* the banks.

When the system *does* work, loans are repaid and it encourages economic growth. Money has been successfully created and enters the economy. But banking policy can't be allowed to surreptitiously promote a growth economy. Therefore, the right to *create* money should be a power of the government, not a power of the banks. Banks could still make loans but only loan out money that they already have on hand from deposits, charging borrowers a higher interest rate than they are offering depositors. It isn't clear whether increasing banking reserve requirements to 100 percent (or any other percentage) is necessary to restrict the economy from growing.

2. *Ending standards-lowering globalization and re-regulating international commerce* (point 5). "Free trade" agreements today are unfortunately a back door to environmental destruction. If one country strictly reduces or regulates its environmentally dirty production of economic goods but another country doesn't, then what happens if they trade in the international sphere? Obviously, the country that uses cheap dirty coal and

slave labor is likely to be able to produce products much more cheaply than another country that uses only renewables and pays all its workers fairly. But that means that in a free-trade situation, the latter, more virtuous country will sink into debt and the exploitive country will become wealthy and powerful. If countries seek to compete in the world marketplace, they have no choice but to reduce their own standards, for example by burning coal and exploiting workers.

The countries with sustainable environmental policies shouldn't be trading with unsustainable countries in the first place, or at least not without someone paying the price. Tariffs could be placed on goods manufactured in the latter countries to reflect the price their goods would have if they actually paid their workers a living wage and developed clean energy. In other words, environmental (and social justice) issues would override the desirability of free trade. The political climate would need to be *very* different from the current one for this policy to be implemented; now, each country does what it thinks best and there is no international consensus on climate change or any other "sustainability" issue. But we urgently need this discussion; otherwise free trade will undercut any progress on sustainability issues.

3. *Ending neo-colonialism through a multilateral payments clearing union* (point 6). The purpose of a multilateral payments clearing union is to reduce or eliminate trade imbalances (in imports and exports) between countries. This item is more germane to social justice than to limiting the scale of the economy.

Some countries have a trade surplus: they export more than they import. Other countries have a trade deficit, importing more than they export. Generally, countries that have a trade surplus are in an economically stronger position. Today, the countries that have trade surpluses include Germany, China, and South Korea—countries in which the economy is booming. Countries with trade deficits often struggle with a weaker economy.

The difficulty is that this situation is self-reinforcing. Countries with a surplus can easily maintain their superiority, while countries with trade deficits often find them very hard to get out of. There is the risk of a downward spiral of debt. Prosperous countries can use this situation to exploit less prosperous countries by loaning them money and imposing new conditions on loans. This leads to the phenomenon of neo-colonialism: exploitation of a country not by the "old" method of directly occupying the country, but through the financial system. (Daly himself doesn't appear to use the term "neo-colonialism," but others have suggested that international trade exploits poor countries to benefit the rich.[336])

This is happening today with the International Monetary Fund (IMF). The IMF, operating as the agent of business interests, imposes onerous conditions on debtor countries. Debtor countries are forced into such things as lowering wages, slashing public services, or looting of their environmental resources as conditions for further loans. This makes it even harder to get out of debt, and the cycle continues. George Monbiot comments: "The consequences, especially for the poorest indebted countries, have been catastrophic."[337]

In the wake of the Second World War, John Maynard Keynes suggested the idea of a *multilateral payments clearing union*. This would replace existing trade organizations and treaties and require that a country's imports and exports be in balance. If any country was out of balance in either direction—unbalanced toward imports or unbalanced toward exports—it would pay a

---

336 Monbiot, "Clearing up this mess"; see also R. Skidelsky and Joshi, "Keynes, Global Imbalances, and International Monetary Reform, Today."
337 Monbiot, "Clearing up this mess."

penalty. This penalty would provide incentives for nations with a trade deficit to get out of their situation, but also for nations with a trade *surplus* to reduce their surplus by making more investments.

Unfortunately, when this proposal was put forward at the Bretton Woods conference in 1944 (in New Hampshire), the United States vetoed it and replaced it with what eventually became the IMF—in which the US had a veto. The IMF was supposed to help less developed countries, but it didn't turn out that way. The current system encourages fraud, militarist adventures like the war in Iraq, and "big stick" policies to enforce favorable trade relations in which western countries get to pillage the natural resources of less developed countries.

At the time of Bretton Woods, the United States had an ample trade surplus and dominated the post-war world for decades thereafter. Ironically, today the situation has reversed and the United States has a trade *deficit*. But because the US dollar is an important "reserve currency" (it is held by many central banks; many countries, such as China, hold dollars in the form of Treasury bonds), and also because it can veto unpleasant proposals from the less developed countries within the IMF, the United States can escape from its debt more easily by just printing or creating more money.

We should therefore eliminate this neocolonialist game at the beginning, by charging countries a penalty for either a trade deficit *or* a trade surplus.

Conclusions

Major economic changes are needed to bring about a truly sustainable economy. Respect for life and the environment cannot depend merely on personal virtue. It must have the force of law and the support of international agreements. We need to *directly* protect the environment and reverse greenhouse gas emissions, soil erosion, groundwater depletion, and destruction of wilderness, imposing economic and political penalties for depletion of shared natural resources.

CHAPTER TWENTY

FACING FUTURE GENERATIONS
A massive demographic transition

What do we do if we figure out that we need to arrive at a human population of one or two billion—or possibly even less? Does "reducing human population" imply ecofascism, mass starvation, involuntary sterilizations, forced abortions, or mass exterminations?

Any kind of deliberate violence, whether war or starvation, would be highly disruptive of the social order. Moreover, the population that would be reduced in such a scenario would likely come from the least powerful, who aren't consuming that much anyway; such a policy would be ineffective as well as cruel. But then how do we avoid some sort of population apocalypse? Are we doomed?

There is nothing *technically* impossible about nonviolently making a transition to a dramatically smaller economy. And if it is technically possible to do so, then it is our human economic systems, cultural beliefs, and political structures that are standing in the way—exactly what we are discussing here. Our economies, cultures, and political systems may be doomed, but *we* are not.

Ideally, we can pare back population through the same general kinds of policies that we used to limit the scale of the economy, namely economic policies. In fact, it may be that these economic policies will be the *only* ones we need. When couples contemplating a family look at the shrinking economy, they may decide on their own to limit the number of children they have.

Getting there

Before we conclude that we're all doomed, we need to take an inventory of our situation. First, birth rates are already falling everywhere, so it's not even clear to what extent we need to modify what we're doing now. Second, there is no intrinsic *technical* reason why, with birth control, human numbers couldn't be nonviolently reduced to pre-industrial levels. Alan Weisman, author of *Countdown* and *The World Without Us*, says that if the entire world *did* implement a one-child policy, we could reduce our population from more than 7 billion to 1.6 billion within a century.[338] Third, there is no intrinsic reason why an advanced civilization, or several civilizations, couldn't thrive with far smaller numbers. Civilizations in ancient Greece, Rome, India, and China, and even early modern Britain, thrived in a world with less than 10 percent of the world's current human population. There are many aspects of modern technology that would help us immeasurably and yet don't require colossal quantities of resources: things like antibiotics, anesthesia, birth control, basic sanitation, reliable water supplies, books, and universities.

We can likely all agree that we need to stabilize population, but how? This might happen spontaneously, through something

----

338 Jaber, "Feeding the Planet, An Evening with Alan Weisman." Weisman did not, however, endorse the Chinese one-child policy. At the time he spoke, human population was closer to seven billion rather than the current eight billion.

like a "demographic transition"; alternatively, we might need some sort of social inducement, like the Chinese "one-child" policy of the late twentieth and early twenty-first centuries.

The demographic transition

One of the most common ideas surrounding population control centers around the notion of the *demographic transition*. This involves a transition from the stable demographics of high birth rates and high death rates prior to the Industrial Revolution, to low birth rates and low death rates in industrially advanced countries. Both demographic states are more or less stable, with birth and death rates approximately in balance. What is *not* stable is a society in the process of development—when, as was true in many countries in the nineteenth and twentieth centuries, death rates came down rapidly, but birth rates remained high for a while. Once countries became industrially advanced, birth rates came down and population stabilized.

Some writers have argued that a decline in world population is already inevitable, and it is occurring in developing countries as well as developed countries due to growing urbanization and education of women. Two conservative writers suggest that "everywhere, virtually without exception, birth rates are coming down. Nowhere are they going up."[339] These writers aren't environmentalists; they dismiss Malthus, *The Limits to Growth*, and *The Population Bomb*, and they view declining population as a *problem*; whereas from a sustainability point of view, declining population (at least for a while) is *desirable*.

Why is this happening? Some believe prosperity lowers birth rates. In that case, one might say there's an obvious solution to the population crisis: increase prosperity through economic expansion! Reversing population growth will happen naturally as a result of improved living conditions, and we can stabilize

339 Bricker and Ibbitson, *Empty Planet*, chapter 2.

population at a lower level. About half of the world, including China, Brazil, Russia, Japan, and all of Europe and North America, already has sub-replacement levels of fertility.[340]

There are two problems with this approach. First, increasing prosperity through economic expansion isn't going to work in a world of resource limits. Herman Daly asks: "Specifically, if Indian fertility is to fall to the Swedish level, must Indian per capita possession of artifacts (standard of living) rise to the Swedish level?"[341] A worldwide demographic transition based on economic growth would be environmental suicide even if it were possible.

Second, it's not the mechanical application of "rising prosperity" to a country that causes falling birth rates, but rather specific items that are typically associated with rising prosperity. It is generally agreed that increases in urbanization (because children are typically most "economically useful" in rural situations), the status and education of women, and the availability of birth control all contribute to falling birth rates. Increases in the costs of having children have much the same effect, something we see in otherwise prosperous countries as resource constraints start to bite. Sometimes, as in the Ukraine (well before the 2022 Russian invasion), we see *falling* prosperity and falling birth rates, as an urbanized and educated population encounters economic difficulties.

The factors that tend to depress birth rates aren't mechanically linked to economic growth. There are many things that could be done to further the status and education of women, provide birth control, and promote urbanization without straining natural resources. And in the normal course of events, with resource shortages and economic policies to shrink the economy,

340 United Nations Department of Economic and Social Affairs, Population Division, *World Population Prospects*.
341 H. Daly, "A Population Perspective on the Steady State Economy."

children will become more expensive to raise. So we already have powerful forces at work to gradually reduce human population.

Immigration

Herman Daly suggests that we limit immigration so that "births plus in-migrants equals deaths plus out-migrants." This is something of a paradox; we normally associate concern for the environment with the political left but demands for restricting immigration with the political right, so here we have a left-wing goal advanced through right-wing policies.

This is a trickier issue than it first appears. It seems that restricting immigration would limit *national* population as opposed to world population, and what we need is a reduction in *world* population. Immigration should neither increase nor decrease population. But since our ultimate objective is to preserve the environment, it depends on how much a hypothetical immigrant would consume, and this is what Daly is likely getting at. If immigrants to the United States (from low-consumption countries) are going to increase their resource consumption after migrating here, this is a potential problem. Therefore, on *strictly* environmental grounds we should oppose net immigration.

But this is complicated by social justice and humanitarian considerations. Generally, we assume we want to help people in need and oppose racists and racist policies; we would like to ally with Hispanics, African Americans, Asian refugees, and so forth. Much of the opposition to immigration comes from people who aren't at all concerned about the environment; they just don't like non-white racial groups. Does whatever environmental good we achieve by restricting immigration compensate for the political and human damage incurred by supporting racism? Would this even diminish resource consumption, or would it just use an environmental excuse to perpetuate an economy that is already highly unequal and highly consumptive?

We could start by looking at how enlightened (or unenlightened) the environmental and economic policies of the United States are in this hypothetical situation. We could visualize an environmentalist United States with policies that establish absolute upper limits on national resource consumption. In that case, immigrants wouldn't increase resource consumption at all. Total US resource consumption would remain constant, with the immigrant getting a small slice of the resource pie. An immigrant would neither increase nor diminish the American resource pie but would slightly decrease the *per-capita* slice that each American gets.

Such an altruistic policy is unlikely in our currently self-centered American culture, where it would likely fuel anti-immigrant sentiments. But Americans could in such a scenario accept immigrants as a humanitarian gesture, understanding that they would slightly decrease the available resources for all other Americans. The political right often makes the argument that this is happening *now* ("immigrants are taking away our jobs"). Willingness to accept refugees might then become a term that the United States could offer in any international climate or resource treaties. On strictly environmental grounds, either a pro-immigration or an anti-immigration argument is possible, but we need to look at the situation as a whole and be clear about what exactly is taking place.

The one-child policy

Since birth rates are falling everywhere already, it's not clear to what extent we need additional policies. Improving basic human services, educating and improving the status of women, and providing basic security for everyone, when added to rising childcare costs, may be all that it takes. But suppose the natural fall in birth rates isn't enough? It's worth making a few comments about one key large-scale effort to promote birth control: China's one-child policy.

China's one-child policy lasted from 1980 until the beginning of 2016. It had the support of most Chinese while it was in place.[342] Basically, the policy worked, although enforcement sometimes generated notable outrages. There were instances of forced abortions and forced sterilizations, depending on the local authorities' attitudes toward the matter. Beyond that, though, there were three conspicuous social problems created by the one-child policy.

1. The policy led to a gender imbalance in society, with more males than females, due to a cultural preference for sons over daughters. Sometimes the ability to identify the gender of the fetus led to selective abortions to ensure the "one child" would be a son; at other times, daughters were put up for adoption outside of China. The resulting gender imbalance in society meant that many unmarried men were unable to find suitable marriage partners.

2. The one-child policy created numerous one-child families, thus many of the children in the society were "only children" who had no brothers or sisters. Some feel this makes children spoiled. Suicides among parents who lost their only child were also a problem.[343]

3. Caring for the elderly became more difficult. In traditional Chinese society (and the Chinese are hardly unique in this respect), children are expected to care for aging parents. With the one-child policy, China now has a large aging population and fewer children to support them.

We could circumvent some of these problems with some common-sense measures. Communal childcare could allow children to interact with other children. Society could provide

342 Pew Research Center, "The Chinese Celebrate Their Roaring Economy."
343 Attanasio, "China 'One Child' Policy."

support for the elderly so the entire burden doesn't fall on their children. Instead of limiting the allowed number of children to one, we could set up tradeable child allowances in a cap–auction–trade system. Women would receive a set number of allowances to give birth.[344] These allowances would be tradeable, so that women could elect not to have children at all and receive compensation. In fact, under this system it is theoretically possible that the *median* number of children per woman could conceivably be *zero,* because a majority of women would elect not to have any children and in effect "sell" their allowance to have a child, while a minority of families, by contrast, would try to buy such allowances and choose to have two, three, or even more children, with everything averaging out to one child per family. If too many families are choosing males over females, we could provide credits for raising female children to bring society's male-to-female ratio back into balance. Ideally, there should be some families with two, three, four, or more children, not just families with one child. Violations would be considered, at least the first time, as a civil economic violation, with parents "fined" by having to purchase an additional child allowance at the going rate.

This also presupposes that we have dealt with problems of social inequality. Otherwise, we are simply privileging the rich (or those with power, probably white people), who could buy more child allowances, over the poor. Even if we could build a consensus around such policies, we would have new social problems to deal with. It's not hard to imagine that people would sometimes act irrationally with respect to decisions to have unprotected sexual relations! And even when the policy works perfectly, it still implies that some adults (perhaps most) will have no children. For these people there need to be alternatives to the basic social structure of the family.

....
344 de la Croix and Gosseries, "Population Policy through Tradable Procreation Entitlements."

Nothing we do will be completely satisfactory, but it pales in comparison with the prospect of a destroyed planet.

Conclusions

In the coming century or so, we should be able to achieve substantial reductions in population nonviolently. A substantial portion of this reduction will come spontaneously, as couples survey the economic scene and rethink their family planning options; and this is already happening world-wide. If this doesn't happen fast enough, we can resort to social policies to facilitate this demographic shift. Despite its flaws, China's one-child policy demonstrated that such policies will work. With the benefit of hindsight, we could make substantial improvements on that policy.

Because both problems and solutions to the environmental crisis increase and decrease with human population, establishing sustainable population levels can transform unsolvable problems into easily manageable ones. Reduced human numbers would alleviate the pressure on declining natural resources and allow humans time to establish a steady-state economy at a lower level of human population and resource consumption.

CHAPTER TWENTY-ONE

RESISTANCE, WORK, AND TECHNOLOGY

Social justice in a post-consumer world

How does the problem of protecting the environment interact with principles of economic justice? If we try to limit economic growth, won't this result in a permanent economic depression, accompanied by poverty and unemployment?

This is one of the most obvious and stickiest problems associated with limits to growth. It is difficult enough to ask people to make what many will regard as sacrifices. Asking for sacrifices in the context of blatant inequality will likely make the problem unsolvable. Restrictions on burning coal would protect the climate but will inevitably mean that some coal miners will lose their jobs. Perhaps the coal miner out of a job can find different work, but if we are trying to scale down the economy, will there really be other jobs?

The quick answer is that we need an immense redistribution of wealth. There should be a minimum income and a maximum income (point 3 in Daly's ten points). Rich nations, and rich individuals in each nation, must bear the primary burden for dealing with the end of economic growth. It's the rich who are consuming the most resources to begin with. In the meantime, we need to make it as comfortable and secure as possible to live a simple lifestyle toward the bottom of the current economic order.

Social justice issues and the environment

Economic injustice is widespread in the United States and the world today. Economic gains are increasingly funneled to the richest "one percent" of the population. One sociologist recently concluded about the United States, "... the concentration of total household income is extremely high ... there is little question that members of the one percent are extraordinarily privileged."[345] On top of that, the United States now jails more people than any other country in the world, both in absolute numbers (more than two million) and in incarceration rates (more than four times the rate for the whole world).[346]

Inequality is widespread throughout the rest of the world as well. *Billions* of people live on less than $2.50 a day, including 44 percent of the population of South Asia who live on $1.25–$2.50 per day. Furthermore, 80 percent of the world's people lack comprehensive social protections such as unemployment insurance and pensions.[347] Moreover, unequal societies are bad for *everyone* in them—the rich and middle class as well as the poor.[348] Social inequality goes hand in hand with a decline of social cooperation and trust,[349] and we will need

345 Keister, "The One Percent."
346 Fair and Walmsley, *World Prison Population List*, 13th edition. The incarceration rate for the United States is 629 per 100,000; the rate for the whole world is 140 per 100,000.
347 United Nations Development Program. "Human Development Report 2014," pp. 3, 19.
348 Wilkenson and Pickett, *The Spirit Level*, makes this argument.
349 See Turchin, "A History of the Near Future."

all the cooperation and trust we can muster to deal with the environmental crisis.

For these kinds of reasons, ecological economist Kate Raworth has proposed that we re-envision our idea of the proper size of the economy as a doughnut (see Box 21-1, "The doughnut economy"). The size of the economy can be thought of as two circles. The outer circle represents the *maximum* size the economy can be to avoid ecological overshoot. The inner circle represents the *minimum* size the economy needs to be to meet human needs in a just and fair way. The "doughnut," the area between the minimum and the maximum size, represents the ideal economy—one able to provide *both* ecological sustainability and social justice.

Addressing environmental issues without addressing social justice could make social inequality worse. People are unlikely to warm up to radical environmental action if they feel that the environment is being protected at *their* expense. The "Yellow Vest" protests in France in 2018 illustrate this kind of problem.[350] As an environmental reform, France imposed a "green" tax on fuel consumption. Protestors felt that this policy fell unfairly on those in rural areas who had few transportation alternatives and had to drive cars and trucks, and they were joined in protest by others who felt that their standard of living was declining.

This is hardly the only time we've seen this type of conflict. In 1990, the northern spotted owl was put on the endangered species list, protecting its habitat from logging. Loggers in the Pacific Northwest protested because they feared, with some justification, that they would lose their jobs. Former President Trump sounded a similar theme in his 2016 campaign when he promised to promote coal as a means to create jobs, which would obviously worsen climate change.

....

350 Cigainero, "Who Are France's Yellow Vest Protesters?"

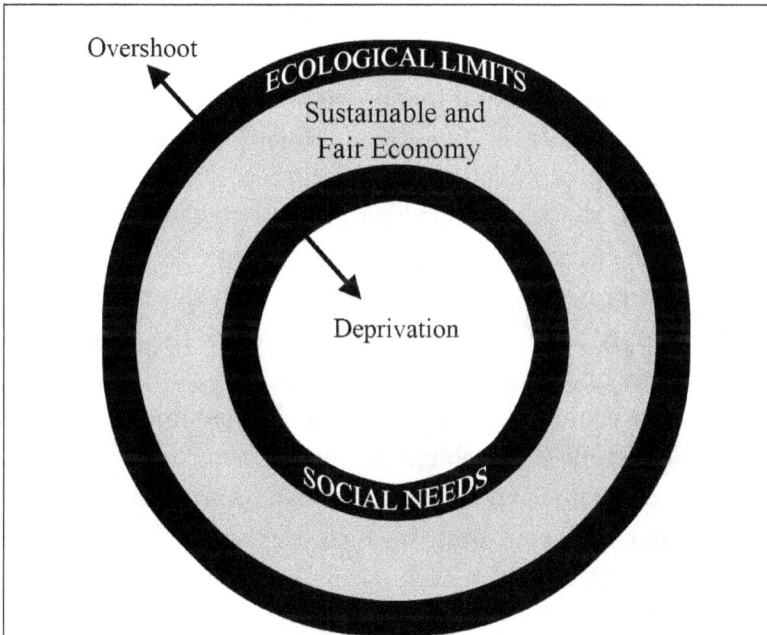

Box 21-1. The "doughnut economy"

Based on Raworth, *Doughnut Economics*, p. 9. The ideal economy shouldn't exceed ecological limits (the outer ring of the "doughnut") but must be large enough to meet human needs (the inner ring). The reality, as Raworth points out, is more complicated than this diagram implies because the economy has already exceeded the limits on some boundaries (e.g., wildlife destruction and climate change), but hasn't exceeded other boundaries (e.g., supplies of common minerals such as iron). But the basic concept is valid for each boundary.

Historically, economic growth is associated with resource consumption. The most recent period of sustained prosperity in the United States began with FDR's New Deal, beginning in the 1930s and extending roughly for forty years. This period, however, coincided almost exactly with the rise of oil as an energy source. This new, convenient, and relatively inexpensive energy source enabled humans to do amazing new things. The economy grew by leaps and bounds, benefiting both the rich and the poor.

Trying to revive economic growth through a "Green New Deal" or some other form of "green growth," though, isn't likely to work nearly as well as the original New Deal (or at all).[351] Unlike in the 1930s, we don't have a cheap new form of energy to exploit; the *lack* of such cheap resources is exactly the problem of limits to growth. In the long run, the economy will have to shrink, simply because the resources on which the economy is based are shrinking.

Since we cannot fix inequality through economic growth, we need to confront the problem of inequality *directly*—through a radical redistribution of wealth. Such a redistribution would be like that endorsed both by Karl Marx and the primitive Christian church: "From each according to his ability, and to each according to his needs!"[352]

Getting to economic justice

The two goals of protecting the environment and reducing social inequality are not that far apart if we think in terms of stopping the wealthy's overconsumption rather than "raising" the poor to the same destructive standard of living. The rich consume far more resources than others. The richest 10 percent (in income), worldwide, are responsible for nearly half of all carbon dioxide emissions, while the poorest half of the world is responsible for only 10 percent of all such emissions.[353]

What this means in practical terms is that the task of restricting resource consumption isn't *intrinsically* that difficult at all. To a large degree it is a social problem. Just by eliminating

351 Hickel and Kallis, "Is Green Growth Possible?"
352 Marx, *Critique of the Gotha Program*, part I. Compare to Acts 4:32–35.
353 Oxfam, "Extreme Carbon Inequality."

the CO_2 emissions of the wealthiest top 10 percent of the population, we would eliminate about half of all CO_2 emissions! (See Box 21-2, "CO_2 emissions and income"). The corollary to this is that if we are going to restrict consumption, this measure will inevitably fall primarily on the rich. The rich can afford substantial reductions in wealth and still enjoy a comfortable lifestyle; much of the poor and lower middle class (or "working class") could be pushed into homelessness with just modest restrictions—and they aren't consuming nearly as much to begin with.

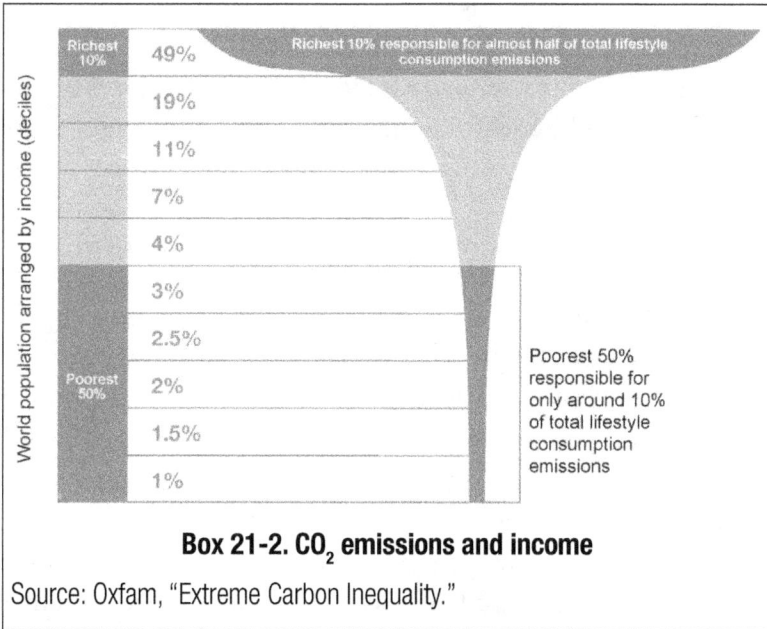

Box 21-2. CO_2 emissions and income

Source: Oxfam, "Extreme Carbon Inequality."

To protect the environment and ensure social justice, we need *two* types of policy, not just one.[354] In the case of the fuel taxes that the "Yellow Vest" protesters despised, the government could put a tax on fuel consumption as planned, but also redistribute income so that the increased income offsets the impact of this tax on the farmers. The rural farmers would still pay the tax,

354 H. Daly and Farley, *Ecological Economics*, p. 360.

but the net impact on them would be neutral or positive. Thus, the rural farmers would have a financial incentive to figure out a way to spend less on gasoline, as that would further increase their effective income.

There are a number of ways we could redistribute wealth. A steep progressive income tax (such as that in the 1950s in the United States) could do the trick. We could implement a tax on *total wealth*, as some European countries have done; if that is too difficult, perhaps a stiff inheritance tax for wealthy estates might eventually accomplish the same thing. We could mandate longer vacations (as in some European countries) and some form of universal health care (something like "Medicare for all" in the United States), paid for by increasing taxes on the rich. This would benefit those at the lower end of the income spectrum the most.

Technology, unemployment, and consumerism

Any redistribution of wealth will have to grapple with the problem of technology. Technology is driving much of the increase in wealth, but it is also—in our current economic system—simultaneously increasing inequality. In the United States, much of the unemployment problem is due directly or indirectly to technology. The *overall* impact of technology, while it does create some new jobs, is to reduce the human workload to achieve a given result. If this didn't happen, we wouldn't develop the technology in the first place. In principle, this is great news, because it means there is less work to be done to sustain our economic output. Unfortunately, it also means that there are fewer jobs to go around, so the rewards for producing this economic output aren't evenly distributed.

This technology and unemployment conundrum has been well understood for more than a century. Many economists thought it would bring about the next phase of social development. John Maynard Keynes postulated that decreasing labor needs would mean that, after achieving a certain desired level of consumption, we would begin to reduce working hours. By contrast, a more skeptical Marx thought that advanced capitalism would lead to unemployment, overproduction, and a falling rate of profit, as the capitalist class utilized fewer and fewer laborers to exploit for their profits. This would lead eventually to a workers' revolution and the next phase of human existence.

What has happened has conformed neither to Keynes's idea of a shortened work week, nor (yet) to Marx's prediction of chronic unemployment and revolution. What has happened instead is that we have gotten on an economic treadmill of rising consumer demand. *Consumerism* has become the solution to technological unemployment. We can drum up consumer demand through advertising; increased consumer demand, in turn, will require jobs to fill this demand.[355]

Dealing with unemployment by creating more jobs through consumerism doesn't really make sense today. The intent of job creation is good insofar as it increases people's *income*. The problem is that we don't really need or want the goods these people will create in their jobs. The United States is already awash in consumer items, and making these goods implies more resource depletion and environmental degradation. "Service jobs" that don't intrinsically require much resource consumption (e.g., college professors or massage therapists) are a little better, but while the jobs themselves don't consume many resources, service workers spend income to buy more consumer items, indirectly promoting consumerism.

Modern capitalism has created a vicious cycle of technology, unemployment, and consumerism. The curse of structural

---

355 See Skidelsky, E., "Unconditional Basic Income and Degrowth keynote lecture."

unemployment due to technological advances has been cured only through the creation of more jobs through consumerism. We are given the choice between unemployment and environmental destruction.

The technology paradox

Technology creates a curious paradox. On the one hand, technology means that many jobs can be automated—thus, in principle, less work for everyone. A future ecosocialist society could then enable us to lead low-consumption lifestyles of relative leisure. The "good life," thanks to technology, would be passing the time with friends and family, reading, volunteer activities, and enjoying the outdoors, while machines did the heavy lifting.

On the other hand, technology—especially information technology—is quite resource-intensive. Computers require all manner of exotic rare earth minerals and mining of scarce and often toxic metals. Resource shortages, especially of the metals and energy required to manufacture this technological infrastructure, means that the resources to support this technology may not always be here.

Do we have—even in a smaller economy—the metals, minerals, and energy to support a society in which information technology plays a critical role in substantially reducing the human labor needed to support civilization? Or are we going to have, in another few decades or another century or two, a fresh resource crisis because we can't afford to support the information technology sector?

If information technology proves to be an unsustainable luxury, that takes us in a *very* different direction. We would need to revert to a more primitive but more sustainable information technology, perhaps that of the nineteenth century (telegraph, books, newspapers, universities) or the mid-twentieth century (radio, television, phones). On the plus side, though, technological unemployment might be largely solved!

Information processing, communication, and coordination are really humans' "super-power" when we think about this problem in evolutionary terms. It would likely be a good thing to preserve as much of the information that we have gained in the past century or two as possible. We would argue, therefore, that we should prioritize information technology as much as possible on a "need to know" basis. There are numerous ways we could substantially shrink the ecological footprint of information technology. This leaner information technology sector could then be propagated throughout the world. Here are some relevant ways we could economize on the resource demands of information technology:

1. The economy will be smaller, and there will be fewer people.

2. Much information technology today is devoted to entertainment. In the meantime, older low-tech forms of entertainment, such as local sports and live performances of music or plays, have gone into decline. It's likely that we could safely reduce the use of technology to provide entertainment and still keep humans supplied with movies and other diversions. Humans have been able to keep themselves amused for tens of thousands of years.

3. Much information technology is devoted to advertising (think of both Google and Facebook). We could reduce the use of technology for advertising. The decline of advertising shouldn't be a crisis; it's in fact a *desirable* outcome. Some advertising is probably necessary for the market to operate optimally, but the current saturation of advertising (and resulting consumerism) into our daily lives has gone far beyond what is reasonably needed.

4. Information technology could be further pared down functionally (perhaps email but no internet), or by sharing information resources (as with computer terminals in public libraries today), or perhaps by restricting access to those with a need to use it (scientists, government officials, businesses, people willing to pay for it). Restricting access would need to be done carefully and democratically, as we have all seen the abuses that are possible here, but if done strictly to conserve resources, it is a possible strategy.

5. We don't need information technology in cars (which will be much rarer anyway), or in household appliances such as refrigerators or washing machines (which we should be grateful just to have in the first place). We'd need to weigh the trade-off in improved efficiency for these devices versus the rarity of the minerals involved.

6. It may be possible to build computers without the bewildering diversity of the rare earth minerals we have today, using more common metals.

For the time being, we will cheerfully assume that advanced information technology will somehow continue to exist in some form and will continue to replace human labor. This brings us to a major proposal to address technological unemployment: the basic income.

The idea of the basic income

A more radical redistribution of wealth is possible: encouraging people toward a lower income through what has been called "the basic income." The basic income is an annual income that would be granted to all adult citizens, regardless of whether they have a job or are even looking for a job. The basic income might not completely solve the problem of homelessness (some people are homeless due to mental illness or other disability), but it instantly addresses that problem as well as a whole host of other social support programs.

The most interesting form of the basic income is one in which is both sufficient to support one person and is unconditional. Some have proposed variations on this policy that impose some conditions, such as proposals for "universal basic services" or for a "government job guarantee."[356] While the basic income is usually advocated for the more developed industrial countries, it might be possible for less developed countries to offer a basic income as well. Unconditional cash payments work quite well at alleviating poverty in less developed countries.[357]

Where's the money coming from for all this? In its simplest form, this isn't a true "government spending" program. It is *both* an income-redistribution mechanism *and* a social-welfare mechanism. It would be combined with a steep progressive income tax (or other progressive taxes), so that taxes will be raised on the rich and given to the poor. But because it is enough to live on, it also replaces many social-welfare programs, decreasing or eliminating the need for such things as disability income, unemployment insurance, and SNAP benefits (formerly "food stamps"). However, in principle, money for the basic income could come from other taxes besides income taxes— say, property taxes, taxes on carbon, or taxes on other natural resources.

Many writers and analysts have recently accelerated the discussion of the basic income, including Andrew Yang, a 2020 US presidential candidate.[358] There is no single basic income proposal, and some proponents are motivated simply by social justice considerations, independent of environmental or

----

356 On the government job guarantee, see Hail, "Paying for a Green New Deal." On universal basic services, see Coote and Percy, *The Case for Universal Basic Services.*
357 Lowrey, *Give People Money.*
358 Bizarro, "United States: Andrew Yang is running for President."

resource concerns. This book's interest in the basic income is that it can meet the demands of social justice without necessarily furthering consumerism or resource consumption. We intend redistribution of wealth through the Robin Hood method: take from the rich and give to the poor. In this way, we want to encourage "simple living" by making it easier to live at the lower end of the income spectrum than at the higher.

Wouldn't an unconditional basic income destroy any incentive to work? Wouldn't *everyone* want to simply collect their basic income and just quit working? This is highly unlikely. The idea that people are naturally lazy and need to be whipped into a frenzy by the threat of homelessness and starvation is a myth. Keeping busy is a fundamental social instinct. Many retirees—people who are *already* getting a basic income under the *present* system—continue working, even though they have an income and typically have other assets saved up over time. Other retirees do volunteer work. In fact, this is extremely common.

Here are some advantages of such a system:

1. A steep progressive income tax plus an absolute floor for adult citizens makes lower-paying but qualitatively satisfying work more desirable and competitive with current low-end jobs. Lower-paying work might be:

    a. A part-time job that allows for more free time; or

    b. Full-time but low-paying work that is intrinsically satisfying (e.g., musician, activist); or

    c. Working unpaid jobs. This includes volunteer jobs and much of what used to be spoken of as "women's work" (and to a great extent still is), like housework and taking care of children.

2. It boosts the security even of people with regular jobs. Everyone would understand that losing a job wouldn't mean workers become homeless and out on the street.

3. It allows us to eliminate or reduce many of the current welfare provisions of our economy. The elimination of much of the complex bureaucracy to administer the system, and the money saved by this process, would offset part of the cost of the basic income. This includes not just the costs of administering welfare, but much of the financial and political costs of propping up the expensive, racist, and unjust prison system.

4. It would deal with the problem of structural unemployment due to automation and computers. Automation would be welcomed because it would mean that society would still reap the benefits of a decreased workload without the trade-off of unemployment and loss of income.

5. The jobs quickest to disappear would be the low-paying, soul-destroying jobs, such as slaughterhouse workers, many fast-food workers, garment sweatshop workers, and retail clerks. If these jobs continued to exist, they would likely demand a much higher wage.

6. A basic income would make simple living much easier and aid in a program of decommoditization (discussed in the next chapter).

7. If in fact inaction on all the resource issues addressed in this book results in some sort of economic collapse along the lines of the Great Depression of the 1930s, but the government and some basic economic services are left intact, a basic income might be useful in providing a "floor" which would avert mass starvation. An economic collapse would result in immense human suffering, so this is, shall we say, not the optimal way to bring about degrowth, but a basic income might enable us to make the best out of a bad deal.

People formulate basic income proposals for a variety of reasons, and there may be some repercussions of a basic income that require further investigation. What would the impact of a basic income be on population objectives? We don't fully understand what drives population up and down now that birth control is widespread. Might a basic income *expand* the economy, as the poor are more likely to spend their money than the rich? Would a basic income tend to fuel, or to dampen, incentives to develop new technologies we need to build a renewable infrastructure? Most critically, with resource depletion, will the technology that makes all this automation possible be sustainable (discussed earlier in this chapter)? But some variation on the basic income is the simplest way of getting to basic economic justice and would aid us in finding a "prosperous way down."[359]

The environmental politics of social justice

There is another facet to the connection between social justice and the environment. As long as it's necessary to acknowledge at least some minimal standard of living for the lower classes, limits to growth will sooner or later force a reckoning with the basic inequalities in our society and direct social struggles in a more radical direction. Social justice radicals and environmental radicals are potentially allies in forcing society in a more radical direction.

Meeting the needs of working people is surprisingly popular. Even the most conservative political leaders, from Ronald Reagan to Donald Trump, have agreed that we need to "create more jobs." They only differ from liberals on the best ways to do this. At least since President Franklin D. Roosevelt (FDR), the United States has used plans promoting economic growth as a means of redressing inequality. This approach purports to create the economic wealth to raise disadvantaged persons to a better level, without upsetting our long-standing class divisions and creating "controversy."

359 As proposed in Odum and Odum, *A Prosperous Way Down*.

Limits to growth mean that *economic growth, as a solution to economic injustice, will not work.* Creating jobs and growing the economy will just increase natural resource use, requiring more energy and materials, propelling us ever faster toward an environmental apocalypse. Limits to growth force a radicalization of our approach to social inequality. No longer can we claim to solve the problems of poverty and inequality by making the economy bigger and distributing some of the largesse to those disadvantaged by race or class.

A social revolution doesn't assure us of success. We could imagine a socialist revolution that burns every last lump of coal to improve the lot of the poor without ever addressing climate change. But a social revolution is necessary, as continued or worsening inequality erodes social trust and guarantees disaster.

To demand a social revolution and an ecological revolution at the same time is a tall order. Surely people will think, "Slow down, slow down—one revolution at a time, please." This natural resistance means that at this time, we likely don't have a political solution that accommodates the need for both social justice and for an ecological revolution. At the first Earth Day in 1970, perhaps, we could have more persuasively talked about an economy that is both "small and beautiful," as British economist Ernst Schumacher wrote in 1975, in his protest against the materialism and greed of the economy.[360] Or we might have talked about "a prosperous way down," as ecologists Howard and Elisabeth Odum did in proposing a transition to a low-energy economy in their 2001 book.[361] In the 1970s, the anti-materialistic counterculture was in full swing, books like *The Population Bomb* and *Diet for a Small Planet* were best sellers, and environmental legislation was sailing through Congress. We could have had degrowth and enjoyed it too, through low-consumption luxuries like storytelling, music, and dance.

360 See Schumacher, *Small is Beautiful.*
361 Odum and Odum, *A Prosperous Way Down.*

Something like that is still possible, but it may be harder to achieve than it would have been in 1970. We may have to wait until a full-scale disaster is right in front of us.

Conclusions

In the past, we were accustomed to deal with poverty through economic growth. In an era of limits, that strategy isn't going to work. We need to have a deliberate policy of encouraging people to live with less, by redistributing wealth from the rich to the poor to alleviate poverty issues. We can encourage part-time work, occasional employment, and more flexible work schedules. A basic income would make it easier and more secure to occupy the lower end of the income spectrum. Environmental sanity requires economic justice as well.

CHAPTER TWENTY-TWO

WHY SIMPLE LIVING IS COMPLICATED

Decommoditizing the economy

Embracing limits means a world with reduced consumption, at least in the United States and other industrially advanced countries. (Many less developed countries are already there, and the poor there need to *increase* their consumption.) What we want is, in effect, what many people call "simple living," but on a *massive* scale.

The problem is that simple living is needlessly complicated. What makes simple living complicated is *commoditization*, the tendency of the economy to promote commodities to fill human needs. There is nothing intrinsically bad about commoditization. Throughout much of human history and even today, commodities have improved our standard of living. This is how we've developed light bulbs and washing machines.

However, commoditization can result in market failures such as those described in chapter 11. Addressing these failures requires something rather different than restrictions on consumerism or protection of wilderness. What we need, as we will explain below, is *decommoditization.*[362]

The paradox of simple living

People have embraced "simple living," the voluntary reduction of consumption, for thousands of years. In different forms it has been incorporated into all the world's major religions as well as the philosophies of numerous secular figures.

Simple living isn't an utterly foreign cultural concept for Americans or for the modern world. Some early Americans, such as the Quakers and Puritans, deliberately embraced simplicity as part of their spiritual practice. Religious communal movements such as the Shakers, Hutterites, and Amana colonies all embraced simplicity, as did nineteenth-century naturalist and philosopher Henry David Thoreau's book *Walden*. Modern American writers contributed to twentieth-century literature on the subject and often combined simple living with nonviolence: Richard Gregg, Duane Elgin, Scott and Helen Nearing (*Living the Good Life*), and Vicki Robin and Joe Dominguez (*Your Money or Your Life*). It can be found in the modern "Simplicity Institute" in Australia, the various degrowth groups and writers in Europe, Japanese minimalism (exemplified in *Goodbye, Things* by Fumio Sasaki), and various blogs. The wildly popular Japanese author Marie Kondo didn't explicitly say to consume less but emphasizes simplifying one's possessions by only keeping things that "spark joy." Many modern advocates also focus on minimizing one's commitments to the outside world to create more "free time" and maintaining internal focus in one's life via meditation practice and minimizing clutter.[363]

....

362 This chapter, and the analysis of "commoditization," obviously owes a great deal to Manno, *Privileged Goods*.
363 Lawrence, *The Practical Peacemaker.*

Why, then, is it so difficult for even the most committed of us to live truly simply in the United States? Just moving out to the edge of town and trying to grow beans, as Thoreau described in *Walden*, isn't really possible for most of us in the twenty-first century. Our economy doesn't support simpler, less consumptive lifestyles. Shopping often requires buying something in plastic packaging or buying more of an item than you need to get the best price. Train service for long-distance travel (much more efficient than flying or driving a car) is infrequent and inconvenient; in many areas, it isn't even possible. Unless you want to live the homeless lifestyle or are already securely retired, it is hard to avoid owning a car, having a job, and renting or owning a house or apartment of some sort, and this will make a certain level of consumption unavoidable.

When you think about it, it's easy to see why the modern economy has this effect. There is more money to be made in manufacturing cars that require oil than in manufacturing bicycles, buses, or sidewalks, that don't require oil. Consequently, we have a vast network of superhighways but fewer bike paths, trails, sidewalks, or mass transit. As with transportation, so with housing; urban sprawl is now engineered into much of the landscape, so cars turn out to be more required than optional. In turn, people need a good job to support themselves, their car, and their house, and the commoditization of the economic landscape becomes almost complete.

What are commoditization and decommoditization?

Simple living is complicated because of commoditization. To see why commoditization is a problem, it is useful to think about the cases where commoditization is *not* a problem. An entrepreneur sees a need, figures out how to make an object, device, or service that fills that need, then markets it to the public.

The problems arise in the last step: taking a good or service to the public. Goods that fill human needs aren't always the most *marketable*. Something has *high commodity potential* (that is, can be easily marketed) if it is a tangible object, if purchasing it has predictable results, and if it can be mass-produced and mass-marketed. Something has *low commodity potential* if it is an intangible service, if you never know quite what you're going to get, and if it can't be mass-produced.

The low-commodity-potential good may be the more *suitable* alternative, however, in the sense that it actually best meets the human need in a way that the person would recognize. Home-grown tomatoes are almost universally recognized to be superior to commercially-grown tomatoes. Marketability and suitability typically diverge when what best meets a human need is labor-intensive or requires special skills, while what is most marketable is something that is shelf-stable and can be mass-produced and mass-marketed. They may also diverge if a good's marketability lies chiefly in the clever marketing strategy itself rather than any intrinsic need: for example, cheap and rapid production of the latest fashions ("fast fashion").

Children's entertainment illustrates how commoditization operates. Let's suppose you want to keep your young daughter occupied and entertained. There are numerous alternatives. You could buy her a Barbie doll, or you could put her in childcare; you could play with her yourself, or you could arrange for her to get together with the neighbor's kids.

But not all of these options can be easily commoditized. If you're an entrepreneur trying to sell something, the Barbie doll has high commodity potential. It is shelf-stable and uses conventional methods of sales and distribution. Childcare may be what you *need*, but it has lower commodity potential. It can't exactly be mass-marketed and bought off the shelf. For both the consumer and the entrepreneur, childcare is rather unpredictable and labor-intensive. And it would be almost impossible to "market" interpersonal play with the neighbors.

The result? Entrepreneurs will naturally focus research and development on things that *can* be sold off the shelf and are perfectly predictable in their functioning—thus, on Barbie dolls in preference to childcare or interpersonal play. In general, the economy focuses research, development, and marketing efforts on commodities that make money, rather than things that— though they might be nice ideas—won't make money. This so obvious that it hardly seems to be worth mentioning. How could it be otherwise?

This process doesn't always work well. Thus, we see the rise of junk food, bottled water, an unhealthy car culture, fast fashion, and planned obsolescence, even though these things don't really fill human needs. They may give us what we want superficially, but come packaged with other disadvantages: obesity, plastic pollution, thousands killed in auto collisions, clothing that we won't want to wear next year, or kitchen gadgets that break down after six months. Commoditization also has a second unintended effect: the goods with lower commodity potential that we truly need typically become more expensive. That's the problem of commoditization in a nutshell (see Box 22-1, "Examples of commodity potential in goods"). How should we respond to this problem?

Economic sector	*High* commodity potential	*Medium* commodity potential	*Low* commodity potential
Food production	Commercial fertilizers	Stored seeds, research services, Community Supported Agriculture	Knowledge of the soil
Health care	Drugs, insurance, hospital supplies	Doctor services, hands-on treatments	Knowledge of nutrition, lifestyle changes
Transportation	Cars, roads	Public transportation	Bicycles, walking
Mental health	Mind-altering drugs	Life coaches	Friendship, exercise
Finance	Junk bonds, credit cards	Credit unions	Personal loans, gifts
Children's play	Barbie dolls, action movies	Childcare, live entertainment	Group play, interpersonal goods

Box 22-1. Examples of commodity potential in goods

Based loosely on Manno, *Privileged Goods*, Table 2.1, p. 27. Ecological economist Jack Manno has done more than anyone else to focus attention on the problem of commoditization.[364] "Commodity potential," for Manno, means "marketable," so that high-commodity-potential goods are more marketable than low- or medium-commodity-potential goods.

....

364 Manno, *Privileged Goods*.

The decommoditization of transportation

In the industrially advanced countries, social infrastructure and customs make it easy to be consumptive and difficult to live simply. To get anywhere in the United States, most of us need a car that burns gasoline. Electric and hybrid cars are only slightly less consumptive. If you want to be a typical consumer and drive a car to get somewhere, the way is already "paved," both literally and figuratively. There are car dealerships, gas stations, and a vast road infrastructure. The most energy-efficient modes of transportation, such as bicycles, walking, public transportation, and passenger rail, don't get a lot of support from the economy.

The bicycle is not only the most efficient form of *human* transportation—even more efficient than walking—it is also the most efficient form of transportation in *nature*, in terms of calories expended to move a given weight a given distance.[365] No wonder several authors, and many activists, have advocated car-free living.[366]

But our society is neither bicycle-friendly nor friendly to any kind of public transportation. Bicyclists are highly vulnerable in traffic; an accident that would be a minor fender-bender if all parties are driving cars could easily become fatal if one of them is a bicyclist. Bicycling is problematic in bad weather, it doesn't work well if the distances are very far, and without special equipment bicycles cannot carry heavy loads. Public transportation is another less consumptive alternative that also isn't well supported in much of the United States. To use it, you need to plan carefully, pay a fare that typically exceeds the costs of driving, and arrive at your destination later than you would have if you'd driven a car. The occasional exceptions—many people cite New York City or Toronto in Canada—tend to prove the rule.

365 S. S. Wilson, "Bicycle Technology."
366 Alvord, *Divorce Your Car*; and Balish, *How to Live Well Without Owning a Car.*

The layout of most cities seems designed to make walking, bicycling, and public transportation as difficult as possible. Sometimes this outcome was accidental, but sometimes deliberate. With the advent of the car in the twentieth century, cities tended to sprawl outward, so that typically many desirable destinations are located far away, unlike in the nineteenth century.[367] Most housing in America presupposes an automobile and the entire road infrastructure based on fossil fuels. You *can* live simply and ride the bus or use a bicycle, but it tends to be complicated, time-consuming, and sometimes dangerous to do so.

Cars can be easily marketed and the infrastructure to use them is already there. Because of commoditization, the likely response of the economic system to the climate crisis will be to promote electric cars. Unlike bicycles or buses, there is substantial economic reason for private investors to be interested in electric cars. While electric cars—if powered by carbon-free electrical energy—would indeed be less consumptive of fossil fuels, they are only marginally "simpler" than standard cars in terms of consumption of *other* resources. Large amounts of metals and technology are involved in electric cars, compared to buses, trains, bicycles, or walking.

If our society is to commit to making it easier to live simply, we need to address these kinds of logistical issues. Public policy could intervene to change the infrastructure. For example, cities could change zoning and housing codes to encourage locating housing, shops, and businesses closer together to create walkable and bikeable cities. Or cities could provide inexpensive (or free) quality public transportation and bicycling infrastructure such as bike paths.[368]

....

367 Kunstler, *The Geography of Nowhere*, especially chapters 6, 7, and 8.
368 Bicycle Colorado in Colorado, and Living Streets Alliance in Tucson, Arizona, are two examples of groups working for just such changes.

The decommoditization of medical care

Living simply, in terms of medical care, is more than just cutting back on use of current medical infrastructure. We could certainly use more self-reliance when it comes to health issues, but not if it means that people wouldn't go to the doctor for problems that could turn out to be serious, causing their health to suffer.

Some low-commodity-potential investments would dramatically improve health care. Basic nutritional knowledge—enough to distinguish a healthy meal with beans and vegetables from an unhealthy one of soda and a greasy meat patty on a refined-flour bun—would have an immeasurably greater effect on public health than simply making open-heart surgery more affordable. Teaching people how to cook simple meals would also do wonders for public health. For many people—perhaps most—these things would lead to a more healthful diet and other lifestyle changes.

But nutrition education doesn't make a good commodity. It's not impossible to market nutritional knowledge or cooking classes; you can write books, open lifestyle clinics, offer a cooking school, or offer private tutoring or coaching. But except for a few, this is hardly an easy way to make a living. No one has a copyright on nutritional knowledge. These kinds of skills used to be taught in public schools as part of "home economics," but such courses were typically directed at girls, were sometimes thought to be sexist, promoted processed foods and meat, and have faded in popularity. There are already large numbers of cooking classes offered free or online, but it is only a dedicated few health professionals or professional chefs who are doing serious nutritional education. More than that, nutrition education is at cross-purposes with all the food companies selling us junk and the pharmaceutical companies selling us pills to treat the diseases caused by eating the junk.

Even for standard health care, when medical intervention is called for, commoditization often disrupts medical choices. There is endless marketing of statins, for example, to lower cholesterol levels, and even gastric bypass surgery (often referred to as stomach stapling) to remedy morbid obesity. In both these cases, the nutritional option needs to be explored; it is typically more effective and less expensive. But there is more money to be made with drugs and surgery, which are also easier to market.

There are some areas of modern medicine where commoditization is a good thing. The keeping of medical records has greatly improved due to automation. But there are areas where medical care (at least so far!) can't be automated, such as the doctor-patient interview.

Commoditization is a critical reason behind the health care crisis in the United States. The remedy is new public policy. Rather than leaving nutrition education to free enterprise, we need to intervene to make quality nutrition education a priority not just for doctors and nurses (who often know next to nothing about nutrition!), but for everyone through our public schools. Home economics classes should incorporate these sorts of skills, along with regular subjects such as math, history, and English, and should be required for both sexes. Home economics could also be utilized to incorporate practical money-management skills that steer young people away from out-of-control credit spending and debt.

The decommoditization of housing

The physical dimension of shelter is a major part of our resource consumption. Because modern housing emerged during a period in which energy was cheaper than building materials, it was (and often still is) not energy efficient.

In the 1990s, Dr. Wolfgang Feist, a building physicist, developed the "passive house" concept. This is housing that requires little or no external (or "active") energy sources for heating or cooling. It does this through superinsulation and design. The passive house is heavily insulated, seals the house from air leakages, has proper windows (such as triple-paned glass), and uses a heat-recovery ventilator to get fresh air. Such designs would drastically reduce the energy requirements for both residential and commercial buildings; many wouldn't need any heating at all (the heat would come from body heat, appliances, and the sun).[369]

The basic concept of "passive housing" is not modern. Traditional housing often has reflected such techniques, such as turf houses in Iceland and traditional houses in southern China (which require cooling rather than heating).[370] In the 1970s, similar ideas were developed in Saskatchewan, Canada.[371] Such housing in our current economy would be quite difficult to market and slow to spread. But if passive-house technology, or something like it, could be incorporated into new housing standards, the quality of shelter could be improved throughout the world.

Sometimes *social* adaptations can conserve housing resources, such as communal housing, cooperative housing, or co-housing. These adaptations are often intended as alternatives to the traditional family unit, but they also impact housing. These (slightly different) living arrangements involve sharing part of ordinary housing space—and sometimes, domestic functions like cleaning and cooking—with unrelated people. In the past such groups have included communalistic groups ranging from Catholic monasteries to intentional and utopian communities of the nineteenth and twentieth centuries such as the Oneida community, the Shakers, the Hutterites, and The

369 Passipedia, "Passipedia – The Passive House Resource."
370 Passipedia, "The Passive House – historical review."
371 Mike Reynolds, "Saskatchewan: The birthplace of Passive House."

Farm. As such, they really serve two functions: an environmental function through conserving resources,[372] and a social function through providing an alternative to the traditional family unit. At least during our degrowth period, when population will be declining because many will have only one child (or none), such alternatives would help to overcome social alienation.

The problem of cost

One key problem created by commoditization is that it tends to make low-commodity-potential goods more expensive. Consider that we *need* many of these goods and that they tend to be less resource-intensive, and we have yet another "market failure" of modern capitalism.

This stems from a phenomenon known as "Baumol's cost disease," named after the modern economist William Baumol who described it. He suggested that the cost of labor-intensive and skill-intensive occupations such as hospital care, higher education, and the performing arts (concerts, plays, etc.) would become relatively more expensive over time because wages tend to rise about the same in all sectors, whether they are becoming more productive or not.[373] Baumol and his colleagues referred to health care, education, and the arts as "stagnant industries" or industries with "low productivity growth"; but it's clear that they are talking about low-commodity-potential goods, just using different language.

People employed in "stagnant industries" will *tend* to see about the same wage gains as everyone else in the economy, even though the industries themselves aren't really becoming more productive. If wages in these sectors didn't keep up with the rest of the economy, the people they employ would quit (or want to!). But we need them, so we pay what the market will bear (as with

---

372 Ivanova and Büchs, "Implications of shrinking household sizes."
373 Nordhaus, "Baumol's Diseases: a Macroeconomic Perspective."

doctors and nurses) or reduce wages through political means (as with, all too often, unhappy public school teachers).

It's intrinsically difficult to try to make these industries more "productive." It takes the same number of musicians to perform a Mozart string quartet today as it did in the eighteenth century. One doctor can only treat one patient at a time. One teacher can teach multiple students at a time, but as classrooms get larger and larger, students with special needs, special gifts, or just idiosyncratic personalities will get overlooked in the rush, degrading "education." Contrast all this with the productivity gains of the computer industry, which has seen phenomenal growth in the past fifty years. We can now hold in our hands a device that is orders of magnitude more powerful than what anyone had in 1972, and the effects of automation are rippling through the economy (see chapter 21).

The reasons for this are complex but come down to the problem of *relative* cost, not absolute cost. College education isn't intrinsically becoming more difficult to provide, nor are teachers and professors becoming fat and lazy. It's just that proportionally, we tend to spend more and more on goods that *can't* easily be commoditized. Regardless of the precise mechanism, we can't doubt the outcome: health care costs rise,[374] college education becomes more expensive, and live concerts become pricier.

Another and less obvious consequence of Baumol's cost disease is that the role of government in the economy is growing. The private sector tends to pick up production of goods that make good commodities—the goods with *high* commodity potential. Private enterprise is by its nature good at providing such commodities. Governments get the leftovers: those things

....

374 Baumol's cost disease isn't the only factor in rising medical costs, as we saw in chapter 17; bad diet also has something to do with it. But the economic factors promoting this bad diet are also significant and are tangled up with the basic realities of nutrition.

such as libraries and education that have low commodity potential and don't quickly turn a profit but are generally recognized to be necessary and beneficial. Because government disproportionately absorbs the costs of these necessary but "stagnant" industries, either directly or through subsidies, governmental presence is already growing.

If we arbitrarily respond by insisting that proportional government size be restricted, via a "balanced budget amendment" or by restricting government expenditures as Colorado has done,[375] the result will be to reduce the availability of these services or to degrade and politicize the end product. This is exactly what we see today in public education, with book bans and declines in teacher salaries. There have been some attempts to increase the "efficiency" of teaching, such as learning videos, online courses, standardized tests, and recorded lecture series. But relying heavily on such methods tends to degrade the quality and experience of the old-fashioned way of doing things. Restricting funds for education means that teachers will be unhappy or that people will have to pay for a decent education for their own children. Private schools have only succeeded because of the rise of "culture wars" and the influence of the ultra-rich in politics, who can afford to send their children to private schools but don't want to pay for education for other people's children. If we want these things to be available to anyone other than the rich, this implies a growing government role.

A decommoditization program will mean an even *greater* role for government.[376] Government will be involved to provide both infrastructure changes and sometimes directly provide those goods that don't make good commodities but are necessary for a simpler lifestyle. For example, government should be involved

375 With the TABOR (Tax Payers Bill of Rights) amendment, passed in 1992.
376 Manno, *Privileged Goods*, chapter 8.

in putting home economics and nutrition education in public schools and in building bicycle paths. These goods will result in less consumption of natural resources and will promote the ability of everyone to live a simpler lifestyle.

Other ways to decommoditize the economy

This same type of decommoditization analysis could be applied in many different areas.

1. We could, through appropriate legislation and economic policy, facilitate the repair of common goods rather than just sending them to the landfill. Such services are rare these days, because it is typically less expensive to trash your old equipment and buy a new replacement than to repair old equipment—even though the repair option would usually conserve resources.

2. Plant-based diets could be decommoditized by the simple expedient of removing the mammoth subsidies going to the livestock industry and to farmers growing feed crops, which typically disadvantage the growing of more healthful and less resource-intensive foods such as fruits and vegetables.

3. The best way to fill a human need is sometimes not a commodity at all. Today we have a proliferation of daycare centers to take care of children whose parents need to work. In earlier times, this was considered "women's work" that mothers would provide for "free," so childcare only rarely entered the economic system at all. This isn't to say that women must be stuck with this assignment; either parent (or grandparents, uncles, or other relatives) could also take care of children while Mom is elsewhere.

4.   In today's commoditized economy, pharmaceutical industries can get rich developing the next new pill to treat depression or anxiety. But depression and anxiety could be minimized in many other ways rather than through a pill. It isn't hard to imagine many ways in which a "simpler" decommoditized world might reduce stress and promote mental well-being. We could provide some degree of economic security for all, such as a basic income, for those at the lower end of the economic spectrum. Social media could be regulated or restricted; today, unrestricted use of social media promotes depression.[377] Obviously serious mental issues would still typically require professional intervention. But society should provide incentives to research these alternatives; pharmaceutical industries are hardly going to get rich pursuing them.

5.   Related to mental health issues, advertising could be restricted or regulated. The very *purpose* of advertising is to create dissatisfaction, and it is driven by some of the cleverest and most well-researched media enterprises in history. Not surprisingly, much of this advertising has succeeded, and many of us have bought the message as well as the product—so we are dissatisfied and insecure.[378]

Conclusions

We need to live more simply on a massive scale. We need alternatives that will satisfy human needs while minimizing materials and energy consumption. However, alternatives to a consumerist approach are often much less marketable; sometimes they aren't even commodities at all. We need to sort these less marketable approaches, culling out those that are less marketable because they are bad ideas (electric combs? ugly

377 Hunt et al., "No More FOMO."
378 See, e. g., Kilbourne, *Can't buy my love*; Barber, *Consumed*; and Levin and Kilbourne, *So sexy so soon*.

sweaters?) from those that are good ideas that need some sort of support to be effective.

Simple living does require cultural changes, but it also requires collective action to build out needed infrastructure and provide less marketable but useful and resource-efficient alternatives to everyone. It isn't enough to be individually frugal; we must work together to make simple living as attractive and available as possible for everyone.

CHAPTER TWENTY-THREE

A NEW AXIAL AGE

The shape and character of massive social change

Every time we look at the nature of an ecological civilization from a different angle, humanity's list of "things to do" seems to get longer and longer, and the task of getting there seems more and more impossible. We need to devote half of the planet to wilderness; we need "simple living," but on a massive scale; we need to eat plant-based diets; we need to lower population levels; we need to redistribute wealth to end social injustice. This already implies completely changing our politics and culture and probably a series of worldwide social revolutions on top of that. How will any of this be possible, when more than half of our current Republican congressional delegation in the US doesn't even acknowledge that human-caused climate change is an actual thing?[379]

Hand in hand with these social and political changes, though, we also need a change in *culture*. Culture, in this context, refers to things that don't really need to be legislated or even

----

379 Drennen and Hardin, "Climate Deniers in the 117th Congress."

stated outright. They are habits and values that are invisible just because they are everywhere. Embracing limits has a cultural dimension in the realm of consumer goods, in the food that we eat, and in the decision to have (or not have) children. Changes in our culture would help us get to an ecological civilization and sustain such a civilization once we got there. But what are these changes and how, exactly, would such changes come about?

What kinds of cultural changes do we need?

To a certain extent, such cultural norms are already here and have been around for thousands of years. It has often been remarked that some equivalent to the command from the Jewish law, "love your neighbor as yourself" (Leviticus 18:19), is found in almost every major religion of the world. At least three specific norms are almost universal: do not kill, do not steal, and do not lie. These are taken for granted by Jesus (Luke 18:18–24), by Buddhists (they are three of the five basic Buddhist precepts), by the other major religions, and by secular philosophies and legal systems from Plato and Epicurus up to the present day.

We already have this sort of widespread basic morality, yet we also have tremendous ecological destruction. The problem is that these principles haven't been applied to the non-human world. Defenders of animals are already familiar with this issue. Protecting species or ecosystems can be done, but it is often an uphill battle with vested interests. Outside of these protections, you often can't stop someone from hurting or killing an animal *unless* the animal is someone's "property"—but even then, only because it is an injury to the property owner, not to the animal itself. The same issue constrains environmentalists trying to protect ecosystems, wilderness areas, or natural resources.

We have the ethical principles already. We now need to somehow extend these ethical principles from humans to the non-human world and from individual virtues to social virtues.

We aren't that far removed, culturally, from such an extension. In the injunction "do not kill," we already have the seed of a more global value of reverence for *all* life. In Buddhism, the first precept—sometimes stated as not to injure any sentient creature—makes this explicit, even though Buddhists don't always consistently apply it in this way. In an ecological civilization, "do not kill" should be extended to animals and indeed to entire ecosystems. We should be moved to protect not only human lives, but also, to the greatest extent practical and possible, the lives of animals and the existence of ecosystems. Even plants and other natural resources that are not sentient and do not "suffer" in the same way humans and animals do, if at all, still need our protection. They are part of the network of life upon which all beings depend and that we should work to preserve.

"Thou shall not kill" should be further strengthened to include not just the command *not to destroy* life, but also a positive injunction to *protect* life and take steps to stop acts of killing.[380] Likewise, "not lying" should extend beyond *not* speaking falsehoods to encompass the *positive* virtue of speaking in a way that decreases suffering. And similarly, while "not stealing" is usually thought of as abstaining from illegal taking of private property, this injunction should also carry the positive implication of *generosity*. In an unjust society, private property could be considered a form of theft.

Social justice is almost certainly a requirement for an ecological civilization, and it too should be expanded to include the natural world. Today we have the theft of wilderness areas used by countless animals, plants, and fungi. On factory farms, we also have the theft of domesticated animals' beaks, tails, sexual organs, offspring, and opportunity for a full lifespan.

---

380 The explanation of the Buddhist precepts by The Plum Village tradition (following Thich Nhat Hanh) makes this more explicit, by suggesting that practitioners learn "ways to protect the lives of people, animals, plants, and minerals."

Think now of our destruction of the lives of billions of innocent animals each year, our driving numerous species extinct, our devastation of entire ecosystems, and our thoughtless pillaging of the earth's natural resources. These tragedies should elicit universal moral outrage, just as much as killing humans today generates such outrage. In an ecological civilization, it *will* elicit moral outrage. If our whole society were re-oriented in this direction, we would have the ethical and cultural understanding needed to address limits to growth and live more simply.

Well-known environmental studies professor David Orr argues[381] that *biophilia* (discussed in chapter 16), our instinctive love or affinity with all life, is innate but can be (and has become) corrupted. It has become a love of nature bent on self-fulfillment—preserving wilderness for sport hunting or scenic vacation destinations— rather than a love of nature tied to a desire to care for nature. We need to *cultivate* a pure biophilia, starting in childhood. To do this requires the participation of the whole community. It also requires access to wilderness and some degree of self-sufficiency in the community; those living in poverty will find it difficult to "appreciate" nature when faced with threats to their basic survival.

To this end, Orr advocates establishment of natural areas, education, and a new covenant with animals: "We need animals, not locked up in zoos, but living free on their terms."[382] This in turn requires new economic and political directions: "The biophilia revolution will also require national and global decisions that permit life-centeredness to flourish at a local scale. ... Real patriotism demands that we weave the competent, patient, and disciplined love of our land into our political life and our political institutions."[383]

---

381 Orr, "Love It or Lose It."
382 Orr, "Love It or Lose It," p. 434.
383 Orr, "Love It or Lose It, pp. 435, 436.

A new ethical understanding doesn't guarantee that we will live in harmony with nature with social justice for all; we still need economic and social policies for that. But a *cultural* understanding of this sort certainly paves the way for such policies and undergirds them.

So, how do we get there?

This is all nice and grand, but how can we facilitate such cultural changes? Are we going to start a new religion, or what? Unfortunately, there is no formula for changing culture or even predicting the future of cultural change; scientific explanations are typically developed only after the fact.[384] We do have notable examples of cultural change in the United States: in the twentieth century, the public became much more likely to accept women and people of color in politics, for example. We could use these movements as a model for similar movements to advance energy conservation, bicycling, economic justice, or plant-based diets.

Doing so does present problems. First, we're after something considerably more ambitious than extending the right to vote. Even those targeted changes took time and haven't fully fulfilled their intent; more than a century has passed, and we're still struggling with racial and sexual inequality. These various movements, present and past, as wide-ranging as they appear, were mostly just modifications within the existing political and social culture, and only aimed at extending rights to other *humans*. Second, the sequence of events and cause-and-effect relationships here are rather murky. Were our various social and political movements the cause or the effect of other underlying technological and social changes in society? Is cultural change something we can consciously bring about, or is it something that just happens?

....

384 "… research on cultural change regardless of its approach has essentially been a postdictive science to date." Varnum and Grossmann, "Cultural Change."

We can find some relatively smaller examples of the sort of change we require. It would help if we could also find documented historical examples of far-ranging cultural changes that resulted in a transformation of society, which in turn altered the course of history across national borders, sending it into a fundamentally new direction. We don't have many such examples; in fact, arguably, we have only one. But it's a doozy.

It's the Axial Age.

The Axial Age

We might think that a universal ethics of "do not kill, do not steal, do not lie" is so self-evident that people would have observed it even in primitive cultures. However, this isn't true. As we saw in chapter 6, primitive hunter-gatherer societies were typically *much* more violent than any modern societies, even during violent wars such as the First and Second World Wars. Obviously, at some point that changed. But when and how?

The term "Axial Age" was coined by the philosopher Karl Jaspers to refer to the period from about 800 BCE to 200 BCE.[385] This time of ferment and new thinking gave rise to many now-ancient religions and philosophies. Through their thought and ideas, Pythagoras, Greek philosophy, the Buddha, the Hebrew prophets, the Hindu sages, Confucius, Lao-Tsu, and many others in the ancient world gave the world a fundamentally different cast.

Jaspers, who initially put forward the concept, didn't offer a lot of explanation on exactly what he meant by the Axial Age. Accordingly, others have criticized the concept of an Axial Age on the grounds of vagueness.[386] What exactly is "Axial" about the Axial Age? What difference did it make in the structure of society or everyday life? After all, there were great leaders who

385 Jaspers, *The origin and goal of history*, Part One, chapters I and V.
386 See Provan, *Convenient Myths*, throughout.

came before this time (Moses, Hammurabi) or after it (Jesus, Mohammed) who expressed the same kinds of ideas as those of the Axial Age thinkers.

Peter Turchin offers two objective criteria for marking out the Axial Age as unique: (1) the Axial Age marks a significant decline in violence between people, and (2) there was a quantitative jump in the size of empires during this time. While we can't be sure of precise dates, *something* truly transformational happened around this time frame.

Primitive societies were generally much more violent compared to historical societies (as we saw in chapter 6). But when did the decline of violence begin? When did primitive peoples cease being "primitive"? Many thousands of years passed between the development of agriculture (c. ten thousand years ago) and the Axial Age and historical empires of ancient China, India, Greece, and Rome (c. 2,500 years ago). It might be *logical* to assume that there was a continuous gradual social and cultural evolution over the past ten thousand years, as Steven Pinker seems inclined to do.[387] But Turchin argues that violence during the past 10,000 years at first *increased*, then started to decrease only during the Axial Age, about 3,000 to 2,500 years ago. It wasn't until the Axial Age, according to Turchin, that we truly began to overcome our violent prehistory.[388]

The Axial Age did more than spawn new philosophers and religious reformers; it was also transformative politically. The Axial Age saw the rise of the first of the mega-empires, the Achaemenid Persian, with 25 to 30 million subjects (beginning 550 BCE). This was followed by the Mauryan Empire in India (322 BCE), China under the Han dynasty (206 BCE), and later

387 Pinker, *The Better Angels of Our Nature*. Pinker doesn't explicitly address the precise timeline of the decline of violence; he seems to believe that it was bad in primitive times, better in the Middle Ages, and best of all today.
388 Turchin, *Ultrasociety*, p. 169, and chapters 9 and 10.

the Roman Empire (27 BCE), each with about 50 to 60 million subjects.[389] Never before the Axial Age had we seen these kinds of large empires.

For Turchin, the rise of these large empires was linked to cultural change and new ethical principles. "The Axial religions introduced several innovations that enabled post-Axial states to increase the scale of social cooperation": constraining rulers and elites to act fairly, thus promoting cooperation; shifting the emphasis from tribal religions to universalistic, multi-ethnic, multi-language societies; increasing trust between people by introducing moral codes that require telling the truth, not killing each other, and so on. Greater cooperation means the possibility of greater empires under a single governance. Thus, the existence of an Axial Age in which morality (as we understand it) gained the upper hand has empirical confirmation.[390]

We might add that the Axial Age also appears to be the time when the idea of a universal ethics achieved *literary* permanence: for example, the prophetic literature in the Bible, early Greek philosophy, and the Analects of Confucius. It's not clear (and we may be unable to determine) whether such literary permanence was a cause or an effect of the spread of cultural and social stability at about the same time, but it was a real phenomenon. Such ethical systems became institutionalized in forms that not only promoted the rise of more stable societies but also survived the rise and fall of the various empires. Buddhism survived the end of the Mauryan Empire, and Christianity survived the end of the Roman Empire; both helped give continuity to the underlying social order. Even though the ancient Indian, Chinese, Greek, Hebrew, and Christian metaphysical systems are quite different, the underlying ethical tendencies do broadly converge. The widespread use of the "Golden Rule" and

---

389 Turchin, *Ultrasociety*, p. 201.
390 Turchin, *Ultrasociety*, pp. 207 ff.

near-universal adoption of the three precepts identified above illustrate these tendencies.

Do we need a new Axial Age to spread a new culture, a new ethics, and a new period of increased social stability? It seems clear, at least, that massive cultural change that promotes social stability is something that *can* happen.

The process of massive cultural change

Well, we may have only one example to go on, but it's worth pursuing how this Axial Age unfolded. There are a few bits of data here worth mentioning:

1. Turchin's research indicates that religion did play an important role in establishing the basic universalistic concepts of the Axial Age, and that moralizing Gods (Gods who establish these ethical principles) do play a role, facilitating the emergence of more complex, less violent societies. However, *the societies became complex first*, then the moralizing Gods appeared.[391] This seems to imply that religion itself isn't a driver of moral change, but that this emerging consciousness (wherever it came from) shaped religion along with everything else.

2. These Axial Age thinkers never attained anything resembling any spectacular worldly success. Confucius and Pythagoras likely thought of themselves as failures in their lifetimes. The Hebrew prophets (Isaiah, Hosea, Amos, etc.) were engaged in struggles that to a certain extent continue to the present day. The Buddha is said to have renounced the kingdom that he would have inherited, and the basic texts of Buddhism weren't even put into writing for several centuries; certainly, the Buddha didn't have any great success during his lifetime. Being an Axial Age prophet wasn't a surefire path to success or wealth.

....

391 Quoted in Wood, "The Next Decade Could be Even Worse."

3.   While the universalistic ethical ideas of the various Axial Age thinkers all tend to converge, the metaphysical systems do not. The five precepts of Buddhism in broad outline closely resemble the ethical principles found in other religions widely separated in space and time. However, the metaphysical systems of various religions are wildly different. Some acknowledge reincarnation; some incline toward monotheism, polytheism, or even atheism; some seem to be uninterested in metaphysics altogether.

Taking all these observations together, it seems that religion was not the driver of these social systems, or at least not *organized* religion. Indeed, as the example of the Buddha, various Hindu sages, and the Hebrew prophets indicate, many of these Axial Age thinkers were in fact protesting *against* the religion of their day, specifically against animal sacrifice (Pythagoras, Hosea, Amos, Buddha). Also, perhaps surprisingly, they functioned as *innovators* in the religious, spiritual, or philosophical worlds; they had no direct immediate impact on politics, although some of them tried and failed at politics themselves (Confucius, Pythagoras).

Primitive societies didn't have this sort of universalistic ethics. In times of stress, even some modern societies seem inclined to revert to a more tribalistic ethics. We shouldn't be dogmatic about such societies or condemn them; they were likely adapting to their "primitive" circumstances. It could be argued that hunting, meat-eating, and warfare made humans what we are today. Perhaps primitive societies evolved to adopt a mistrustful culture in which it is all right to kill unrelated strangers. But without necessarily assigning any moral blame here, the default precepts that we accept today—not to kill, lie, or steal—were not standards of the society in which the Axial Age thinkers arose. What seems to us to be blazingly obvious was *not* obvious three thousand years ago.

A new or an old Axial Age?

Bringing about a new Axial Age is surely an intimidating task. The last Axial Age took six centuries to unfold. On the other hand, the Axial Age thinkers lived in a world of limited literacy and communication across distances. Today we have widespread literacy and instantaneous electronic communication across the entire globe. There is no intrinsic reason why a new Axial Age couldn't occur with—relative to the previous Axial Age— lightning speed.

With the original Axial Age, we reached the possibility of a practical ethics representing all humanity. What we need today is a practical ethical system (or systems) that reflects a deeper regard for *all life on Earth*. We need a morality that includes plants, animals, and ecosystems.

Interestingly, such a morality was *already* present, to a certain extent, in *some* of the Axial Age thinkers, such as Pythagoras, the Buddha, the early Hindu advocates of ahimsa, and some of the Hebrew prophets. It also was present in some Christian groups (e.g., the Ebionites) and Islamic thinkers (e.g., some Sufi mystics).[392] These thinkers expressed respect for refraining from killing animals (and by extension, from harming nature), and their ideas affirm those we have explored here: that one's true family is all life, and that a simple life of minimal consumption is the easiest and happiest one. But rarely has anyone successfully incorporated this extension of "reverence for life" into the daily practice of larger civilizations.[393] It was restricted to isolated thinkers

....

392 Arguably, it is also present in some indigenous peoples' respect for nature, although it is no doubt impossible to generalize across all the different peoples and all historical peoples.
393 India under Asoka may be an exception, and there may be other exceptions as well.

(Pythagoras, Thoreau) or to the monastics in various religions.[394]

The seeds of such a deeper ethical view are *already* both ancient and widespread, and these seeds simply need to be cultivated. The principle of respect and tolerance for *all* life is not springing newborn into our world for the first time, and it still spontaneously arises in modern thinkers throughout the world. In short, these ideas are both old and new. We may not need to start a "new" Axial Age, but only to continue the previous one.

Conclusions

The urge to cooperate is part of our nature and a key to our evolutionary success.[395] But cooperate with whom, and for what purpose? Extending this sphere of compassion to all humanity, and incorporating it into the entire social fabric, is exactly what happened during the Axial Age. We don't have much to go on, except that we know massive cultural change *can* happen. Perhaps today we stand at the beginning, or are even in the middle, of a new such Axial Age, extending compassion to all life on Earth.

---

394 Whether universal respect for life (e. g., teachings on vegetarianism) represented the original teachings of Buddha and Jesus, or was viewed as "heretical" in their respective traditions is an interesting historical question—but not one that matters for our purposes. In any event there is no radically *new* content in a modern ethics of reverence for life.

395 See, e. g., Singer, *The Expanding Circle*; Harari, *Sapiens*; and Turchin, *Ultrasociety*.

CHAPTER TWENTY-FOUR

WHAT DO WE DO NOW?

Activists and ideals

The question "What do we do now?" (as discussed in chapter 3) is ambiguous. Who is this "we"? Is it you and I, the author and the readers of this book, acting alone? Is it some real or potential organization or movement to change the world? Is it the nation, or society as a whole?

During most of this book, when I ask what "we" can do, I mean the nation, society, or the world. "We" can reforest grazing land, we can implement a carbon tax, we can build bike paths, and so forth. But these are all things that only the nation or the world can do; we can't individually build bike paths, tax carbon, or reforest grazing land. Individuals and groups may want to know what to do themselves as well as what they can do to prod nations and the world to "do something"—or perhaps, since most of our economic activity is so destructive, to "do less" or "do nothing."

In this chapter I want to explore an alternative meaning of this "we." "We" are those people who broadly agree with the basic thesis of this book: that humanity needs to take radical action to protect the biosphere. This includes you, if you are generally convinced of the direction of this argument, even if you don't agree on everything I've said or haven't read this book. Of course this "we" is constantly changing, as public awareness changes and as the world situation changes. Answers to these kinds of questions thus become a moving target.

What do we do now?

Taking action

For the most part, "What do we do now?" concerns taking *collective* action. Collective action is bound to be more effective than the lone individual trying to save the world. But any collective action faces several obvious obstacles. First, there just aren't very many of us; so far, at least, "limits to growth" still doesn't seem to register on a scale of national concerns (though our numbers seem to be increasing). Second, there doesn't seem to be much we can do that would have a chance of success. Third, even among this relatively small and relatively enlightened group of people, there is a wide variety of points of view. Some people are doomers[396]; some are apocalyptic environmentalists with libertarian inclinations[397]; others are radical leftists, hoping for social revolution[398]; some include Buddhists who hope for social change through engaged mindful action[399]; others are ecological economists[400]; others are some combination of these or something else altogether.

We do, however, have some assets. The public's awareness of environmental issues is continually increasing. Mother

---

396 E. g., Wallace-Wells, *The Uninhabitable Earth*; Dilworth, *Too Smart for Our Own Good*.

397 Described in Schneider-Mayerson, *Peak Oil*.

398 E. g., Vettese and Pendergrass, *Half-Earth Socialism*; Hickel, *Less is More*.

399 E. g., Nhat Hanh, *Zen and the Art of Saving the Planet*.

400 E. g., Daly and Farley, *Ecological Economics*.

Nature and history are on our side; as humans confront one environmental catastrophe after another, awareness is bound to increase further, environmental issues will start to float to the top of the national psyche, and discussion among those who are aware will promote clarity of vision. But we don't know the exact way in which these events will unfold.

If we want to work together with others, we immediately collide with the problem of dealing with divergent frames of reference. People want to grasp the situation through some kind of "grand narrative" that explains what is happening and why. We need to look at whom else we might be working with before trying to identify more specific actions.

Do we need a grand narrative?

The most likely way that we could bring about something historically momentous would be through a grand narrative that would bind people together in common action. People love stories. Americans are bound together with tales of the American Revolution. Other countries, as well as companies, social groups, and religions, have their own stories. But narratives, in turn, often carry with them explicit or implied strategies. So the job of finding a handy narrative goes hand-in-hand with the job of identifying good strategic choices.

In the innovative book *Degrowth and Strategy*,[401] the authors outline three basic strategies for achieving degrowth:

1. *Ruptural* strategies which seek to smash or attack the existing system, perhaps along the lines of the French or Russian revolutions.

----

401 Nathan Barlow, et al., *Degrowth and Strategy*. They follow the terminology of Erik Olin Wright in *Envisioning Real Utopias*, pp. 303–307.

2. *Symbiotic* strategies, which work within the existing system in order to change it.

3. *Interstitial* strategies, seeking to achieve the goals of degrowth outside of—or in the absence of— the existing system.

Each of these strategic choices invites a different type of grand narrative. During most of this book, I have tried to resist the urge to offer up a grand narrative. Once one offers up a grand narrative, things that don't comport with that story tend to get overlooked or nudged aside. I would prefer to lay out all the evidence as clearly as I can, then see what we can do with the material. The practical problem of persuading people to rally around a cause is important, but first we need to try to understand the reality of the situation in as empirical a way as possible.

When people do become aware of resource limits, they tend to see resource problems in light of whatever pre-existing viewpoints they already held. Thus, libertarians see the folly of large government, socialists see the evils of capitalism, Buddhists see greed and delusion, and so forth. This doesn't mean these narratives are necessarily wrong! It just means we should be cautious about looking at data that seems to conform to what we already believe, especially if we hope to collaborate with people of differing points of view.

Some of you may think that I've already introduced a grand narrative—albeit one that lacks a clear conflict between well-identified good guys and bad guys. If you *insist* on a grand narrative, then my best attempt is in the previous chapter: what we are living through is a second Axial Age, in which the fundamental assumptions of society undergo a complete transformation, enabling the establishment of new (and less violent) forms of society and politics. Or, depending on how you want this narrative to go, it's just a continuation of the *original* Axial Age, which began almost three thousand years ago, and is just now nearing its conclusion—because people didn't fully understand at the time what Buddha, Pythagoras, and Isaiah were talking about.

I can immediately point out several problems with using "a new Axial Age" as our grand narrative. Many different things happened during the original Axial Age. We had Pythagoras and Plato, but we also had the Peloponnesian War. We had Confucius, but we also had the Warring States period in Chinese history. Should we study philosophy and meditate, or study strategy and go to war?

Other grand narratives about the environmental crisis already have some adherents. One alternative narrative is "collapse." Many eloquent writers feel that civilization is doomed and there is not much to do except adjust to the rough world of the future.[402] At best, they offer, we can be *individually* prepared for the future or small local communities might live sustainably, a narrative which suggests interstitial strategies.

I hope that is not the answer. If the only answer to climate change and oil depletion is learning how to grow your own vegetables and getting your health and finances in order, then we have a problem. There's nothing wrong with "prepping"; this author and his wife have a backyard garden and are trying to stay healthy. But we don't want this narrative to deaden thoughts of social action.

Even looking at the problem narrowly, personal survival in the face of disaster will depend overwhelmingly on the survival of our *community*. If your local community is socially cohesive, someone could probably teach you gardening skills and the basics of nutrition. Moreover, there's nothing impossible about governments covering a large territory in a situation of minimal resource consumption. The United States in its early

---

402 See, e. g., Kunstler, *The Long Emergency*; Wallace-Wells, "The Uninhabitable Earth"; Tverberg, "The climate change story is half true"; and Dilworth, *Too Smart for Our Own Good*.

history covered a vast territory with a fraction of today's material consumption; going back further, the Chinese, Indian, and Roman empires lasted for centuries on even less energy than that.

Ecological economists often propose changes in economic policy which encourage a symbiotic grand narrative: working within our current system to reform it. They make proposals such as a universal basic income, increasing the availability of birth control, carbon and meat taxes, and protection and expansion of wilderness areas. While quite radical, these are broadly compatible with our current economic framework.

A third possible grand narrative is social revolution. How likely is it that the very worthy proposals of ecological economists can be realized without a revolution? The social and economic conditions today do broadly resemble earlier social upheavals.[403] The political and cultural changes we have in mind surely require something *like* a social revolution. Calling it a "revolution" brings up images of masses of people demonstrating in the streets, rising up against an oppressive system. Thinking in this way correctly conveys the scope, difficulty, and urgency of the problem and certainly presents a plausible model of action.[404]

The "social revolution" model also has some limitations. Revolutions aren't all peace and love; they often involve long and violent civil wars, so we might want to think twice about this before we dash out into the streets and mount the barricades. Perhaps we could follow Gandhi's example in a "truth-force" revolution, in which Gandhi pursued independence for India using nonviolent means. Another serious problem is that most

....

403 See discussion in Turchin, *Ages of Discord*, and Turchin and Nefedov, *Secular Cycles*.
404 Hickel proposes a "post-capitalist world" in *Less is More*. Kovel (*The Enemy of Nature*), Foster (*The Ecological Revolution*), and Vettese and Pendergrass (*Half-Earth Socialism*) offer socialist or Marxist perspectives.

revolutions focus attention on an oppressed *human* group, with activists working to increase "class consciousness" to unite against their oppressors. Unfortunately, reality is more complicated than that. The history of the former Soviet Union could serve as a warning. The USSR was sincere, at least, in its desire for *human* equality, but that agenda also increased fossil fuel emissions and accelerated soil erosion.

If you think about these various narratives, you can see that they aren't mutually exclusive, nor do any of them exclude the Axial Age narrative. We live in a unique historical period, and it might be best not to dogmatically hang on to any presuppositions about what will, or will not, be required to transition to an ecological civilization.

The example of the Axial Age suggests that the key new element of the Axial Age was not innovation in military or political tactics, but a slow-moving "consciousness revolution." Why should we, today, use violence against the existing order? No amount of disruption that we might inflict can equal what Mother Nature is going to impose on our entire society. So rather than impose some grand narrative on what is—even as I write this—in the process of unfolding, we could just let any ambiguities here stand on their own for the time being.

What can individuals and groups do?

What about more specific actions? Do you feel energized, ready to go out and change the world? The possible actions we could take spread out in a thousand different directions, depending on one's platform, audience, and grand narrative of choice. But the following ideas are compatible with multiple narratives and provide an opportunity for people with divergent views to discuss the various alternatives.

1. *We can reduce our ecological footprint as individuals.* There are some things individuals can do all on their own, outside of political or social activism. Do we have something analogous to *50 Simple Things You Can Do to Save the Earth*?[405] Certainly most of the activities mentioned in that book would help: recycling, composting, energy conservation, and so forth. But these activities, even if everyone did them, don't come close to what the task will require.

We could try to live with Gandhi-like simplicity, but as we found in chapter 22, living simply in the United States is complicated because the infrastructure for our entire society to live simply doesn't yet exist. However, there are two actions that require very little infrastructure that isn't already available and that would have an outsized impact: (1) adopt a plant-based diet and (2) limit the number of natural children you have (ideally to zero or only one). Either of these can be socially painful, and protests from well-meaning relatives and friends may discourage people from going vegan or curbing their reproduction. Of course some of us already have kids, so we can't exactly turn back the clock on that decision, but the decision to adopt a plant-based diet is available to anyone at any time.

2. *We can support existing groups.* Collective action requires finding others with similar beliefs. How do we do this? Everyone's situation and interests are different, so there's no single answer that will work for everyone.

Many of the nuts and bolts of getting to an ecological civilization are already in hand. What we'd like to see is one, or several, major groups with both a national and a local presence incorporating a "limits to growth" perspective similar to that advocated in this book: reduce consumption, reduce population, and drastically reduce or eliminate livestock agriculture. As I write this, there are a few groups that are moving in this

405 Earthworks Group, 1989, 1995.

direction.[406]

There are other groups that, while they don't explicitly adopt a "limits to growth" perspective, in *practice* function (or might function) in that way. Environmental concerns such as climate change, plant-based diets, population, transportation alternatives to the car, and biodiversity *already* have thriving communities of concern. Yet other groups may at least be open to listening; your friendly local church, synagogue, sangha, temple, or secular hub might be interested in hearing or even spreading this message, since many spiritual and secular communities are concerned with issues that touch on the environment.

The bad news is that these communities of concern aren't always talking to each other and are generally unaware of the problem of limits to growth. Many assume that the economy will continue to grow and indeed should grow, just "sustainably." This brings us to our next task:

3. *We can work to increase consciousness of limits.* As I write this, there is approximately zero consciousness of limits to growth! The "anthropocene," the proposed new geological period in which we currently live, and in which humans have exerted a decisive influence over the environment, is still utterly ordinary to so many people in industrialized societies, even though to us it's clearly not natural.[407] Where we are going requires radical changes that are essentially "dead on arrival" in current political discussions.

....

406 Two of which I am a member: The Center for the Advancement of the Steady-State Economy (CASSE) at https://steadystate. org/; and the Center for Biological Diversity at https://www. biologicaldiversity.org/. There are other groups; these are just the ones with which the author is most familiar. The degrowth movement is active in Europe but less so in North America. Also worth mentioning is the Post-Carbon Institute.
407 Swanson, "The Banality of the Anthropocene."

But while there is little consciousness of limits to growth as an issue, there is a large and growing consciousness of *particular* limits. *People see the "trees," but they don't see the forest.* They see *their* environmental or social problem, but they don't see that their problem is just one facet of a bigger problem: that of limits to growth. They see carbon dioxide emissions but don't see factory farms, or they see factory farms but don't see overpopulation.

Many climate change proposals presuppose continued economic growth and fail to factor in resource depletion. Many wildlife conservation efforts are based on making wilderness preservation financially lucrative (as in "wildlife tourism"). Social justice activists may propose jobs programs (thus presupposing that the economy is the salvation of the poor). These aren't necessarily bad ideas! But we have lost sight of the forest in meditating on each individual tree.

We can promote consciousness of limits by explaining to those making such proposals, whenever and wherever we have the opportunity, that we agree with their objectives but feel that to achieve them, we must acknowledge and address limits to growth. It is ludicrous, at this stage, to assume continued economic growth and unlimited resources.

4. *We can promote social justice.* The hostility of American elites to the concept of limits to growth may be, in the long run, the most significant obstacle in dealing with the problem of limits. *Logically,* limits to growth and environmental protection don't seem to have much to do with social justice. But in practice they have *everything* to do with social justice (see chapter 21). Limiting economic growth is going to hurt the elites, and the elites will, generally speaking, oppose any meaningful environmental action. This isn't an iron-clad rule; some of the wealthy are informed, alarmed, and willing to do something both about social justice and the environment. But the physical power

of the elites comes from environmental destruction. Those who oppose the power of the elites for social justice reasons and those who oppose the power of the elites for environmental reasons are potentially allies.

5. *We can support specific actions or legislation.* We should also keep specific proposals in our back pocket, so to speak, so that we could pull them out if the opportunity arises to debate or propose such actions or legislation. A partial list of such proposals might include: cap-and-trade proposals for greenhouse gas emissions; a meat tax and ending subsidies for the livestock industry; ending standards-lowering "free trade" agreements; expanding access to birth control; wilderness protection and expansion, along the lines of the Buffalo Commons; and nutrition education for doctors and the general public. Some of these proposals have recently surfaced in public debate: a cap-and-trade proposal (see chapter 19) briefly came before Congress in 2009, and a universal basic income (see chapter 21) was proposed in the 2020 Democratic presidential primaries.

6. *We can stay engaged with science.* Science is not a static thing. It is a dynamic community of scientists. Science is dependent on the wider community and can be subject to political pressures. Science also can undergo revolutions—just like societies. Today, there are two fields that are under intense revolutionary pressure of a *scientific* nature: economics and nutrition. If economists declare that we have reached the limits to growth, or if doctors and nutritionists agree that plant-based diets are healthier, two critical pieces in the puzzle of embracing limits will gain substantial support.

There are powerful economic interests that would like to manipulate science so that it comes to the conclusions they would *like* it to come to. Failing that, they will manipulate the public image of science to cast doubt where there should really be consensus, and consensus where there should be doubt—or even consensus in the opposite direction. Not everyone will have the opportunity to participate in or to affect these debates, but we should at least be aware of the challenge that ecological economics presents to mainstream economics, and that plant-based nutrition presents to the nutritional mainstream.

7. *We can promote positive cultural changes.* Any set of massive social changes will involve cultural changes of some sort. We don't know exactly how this works or could work. Historians argue that the French Enlightenment (Voltaire, Rousseau, etc.) was a key cultural cause of the French Revolution, and that the eighteenth-century Great Awakening in North America (which led to questioning of religious authority) was a key cultural cause of the American Revolution.

Even as we speak, there are doubtless many cultural changes which are preparing the way for the next steps, as well as groups promoting these changes. What we really want is a change in underlying social mores: not only that we shouldn't oppress or exploit other humans of a different race, sex, or social class, but also that we shouldn't engage in mindless consumerism, consume food that is destructive of animals and the earth, or have more children that will expand the human presence on the globe at the expense of nature.

Waiting for the apocalypse

Dramatic action on the environment seems highly unlikely today. There is political gridlock both within and between nations. Mainstream economists don't acknowledge the problem of limits, and our political leaders rely on these economists. Environmental resources that support the economy are melting away, and a collapse of our economic and political systems seems to be the default outcome.

But we shouldn't give up just as things are beginning to get interesting! Our economy is precisely what is driving all this environmental destruction. Even though collapse is obviously a dangerous situation for humans, we shouldn't *necessarily* fear such a collapse in and of itself. On one hand, it could mean a descent into a more primitive and barbaric social order. Or it could be the entry point into an ecological civilization in which a smaller economy can support a more enlightened social order.

Promoting awareness of environmental problems doesn't appear to be a fundamental problem; Mother Nature is going to help us on that score. Sooner or later, a public epiphany is coming: the shock when we collectively realize the gravity and nature of the environmental crisis.

We might be surprised at the results of such a shock. Charles Fritz, a prominent American sociologist, investigated the role of traumatic disasters on the mental health of the survivors. During the bombing of Great Britain during the Second World War (the "Blitz") the morale of civilians *improved* and mental illness declined. Great Britain at that time was "a nation of gloriously happy people, enjoying life to the fullest."[408] The same thing happened in Germany later in the war, with roles reversed. The Allied bombing of German cities improved German civilian morale and German industrial production *rose* as the bombing intensified. Bombing civilian areas likely delayed the end of the war.

Why did these calamities have the effect of promoting mental health? Fritz postulated that in modern life, the traditional primary-group bonds have been shattered by what we would ordinarily call "progress"—technical advances and economic expansion.[409] We are now prosperous, but alone. We don't need others, but on the other hand, no one needs us, either.

Even though a disaster is a calamity, in terms of the effects on the group it may bring social connection back to life. Fritz observes:

408 Fritz, *Disasters and Mental Health*, p. 4.
409 Fritz, pp. 25 ff.

Disaster provides a form of societal shock, which disrupts habitual, institutionalized patterns of behavior and renders people amenable to social and personal change. The essential effect of shock is to arrest habitual repetitive patterns of behavior and to cause a redefinition and restructuring of the situation in accordance with present realities. ... A shock, therefore, always contains the seeds of change.[410]

A disaster such as economic or political collapse doesn't *guarantee* success. But at least the opportunity for a cultural and political transformation will be there.

Conclusions

The problem of limits is already here. In many ways we've lived beyond these limits for many decades—including in the areas of climate change, fossil fuels, livestock agriculture, human overpopulation, mass extinctions, resource depletion, and the devastation of wild plant and animal life. Instead of trying to evade these resource limitations and prop up our bloated culture and economy "sustainably," we need to embrace these limits and work for a different politics, a different culture, and a different ethics.

Awakening involves both consciousness of these limits and meaningful collective action to deal with them. Individual awakening is important. What we need is collective awakening.

....

410 Fritz, p. 55. Emphasis in original, but in boldface.

ACKNOWLEDGMENTS

I really don't know where to start or stop in acknowledging people who helped me with this book. Many of my ideas evolved in conversations, and since I've been working on the ideas in this book for nearly two decades, many of those people and conversations are now somewhat distant memories.

A number of people who saw early versions of the manuscript either made helpful comments or offered substantial encouragement and support along the way. These include Mary Ann Peters, Ted Struzeski, Andie Cogswell, Jean-François Virey, JoAnn Farb, Drew Hensley, Steve Kaufman, Jonathan Balcombe, Howard Lyman, Vesanto Melina, Sailesh Rao, Andrea Metzger, John Pierre, and Victoria Moran. Thanks also to my editor, Julie Miller, who did a fantastic job in helping to get the manuscript into its final form.

Over the years, I was also assisted by many librarians, especially those at Denver Public Library, who helped me in locating obscure books and papers that I couldn't find online.

There are three people who greatly influenced my ideas, and for whom I am grateful, but have now passed on and never saw the manuscript. My father, Dr. Lawrence Akers, introduced me to the idea of peak oil in the 1970s, long before it was trendy; he helped set up a talk by M. King Hubbert for the first Earth

Day in Oak Ridge. My friend for decades, Charles Barton, Jr., engaged me in endless discussions of energy issues, especially nuclear power. Lisa Shapiro, fearless vegan activist, discussed my ideas about ecological limits with me and strongly encouraged me to publish a book on this subject.

I'm almost certain I've left someone out, so my apologies in advance. Any errors in the book are entirely my own.

My wife, Kate Lawrence, talked with me late into the evening on numerous occasions over the ideas in this book. She repeatedly went over the manuscript and made countless intelligent suggestions. To you, my love, I dedicate this book.

BIBLIOGRAPHY

Abraham, Andrew J., Joe Roman, Christopher E. Doughty. "The sixth R: Revitalizing the natural phosphorus pump." *Science of The Total Environment* 832 (August 1, 2022), 155023, ISSN 0048-9697. https://doi.org/10.1016/j.scitotenv.2022.155023.

Ahmed, Nafeez. "Former BP geologist: peak oil is here and it will 'break economies.'" *The Guardian*, December 23, 2013. https://www.theguardian.com/environment/earth-insight/2013/dec/23/british-petroleum-geologist-peak-oil-break-economy-recession.

Aitken, Hugh G. J., editor. *Did Slavery Pay?: Readings in the Economics of Black Slavery in the United States*. Houghton Mifflin Company, 1971.

Akers, Keith. *A Vegetarian Sourcebook: The Nutrition, Ecology, and Ethics of a Natural Foods Diet*. Vegetarian Press, 1993.

Alpert, Jack. "The earth's sustainable population is below 100 million." Stanford Knowledge Integration Laboratory, August 15, 2009. http://www.skil.org/position_papers_folder/100million.html.

Alvord, Kathryn. *Divorce Your Car: Ending the Love Affair with the Automobile*. New Society Publishers, 2000.

American Society of Civil Engineers. "2017 Infrastructure Report Card." https://www.infrastructurereportcard.org/wp-content/uploads/2017/01/Energy-Final.pdf.

Anderson, Ross. "Welcome to Pleistocene Park." *The Atlantic*, April 2017. https://www.theatlantic.com/magazine/archive/2017/04/pleistocene-park/517779/.

Andrade, Jason, Aneez Mohamed, Jiri Frohlich, and Andrew Ignaszewski. "Ancel Keys and the lipid hypothesis." *British Columbia Medical Journal* 51, no. 2 (March 2009): 66–72. https://bcmj.org/sites/default/files/public/BCMJ_51Vol2_Mar_core_1%20%281%29.pdf.

Anielski, Mark. "The Genuine Progress Indicator – A Principled Approach to Economics." *Encompass*, October–November 1999. https://www.pembina.org/pub/genuine-progress-indicator, accessed October 25, 2021.

Arsenault, Chris. "Only 60 Years of Farming Left If Soil Degradation Continues," Thompson Reuters Foundation, December 5, 2014. https://www.scientificamerican.com/article/only-60-years-of-farming-left-if-soil-degradation-continues/.

Asner, Gregory P., Andrew J. Elmore, Lydia P. Olander, Roberta E. Martin, and A. Thomas Harris. "Grazing systems, ecosystem responses, and global change." *Annu. Rev. Environ. Resour.* 29, no. 1 (July 26, 2004): 261–299. https://doi.org/10.1146/annurev.energy.29.062403.102142.

Attanasio, Cedar. "China 'One Child' Policy: 6 Crazy Problems This Approach Created." *Latin Times*, October 29, 2015. https://www.latintimes.com/china-one-child-policy-6-crazy-problems-approach-created-350601.

Australia's Chief Scientist (website). "Which plants store more carbon in Australia: forests or grasses?" December 1, 2009. https://www.chiefscientist.gov.au/2009/12/which-plants-store-more-carbon-in-australia-forests-or-grasses.

Badgley, Catherine, Jeremy Moghtader, Eileen Quintero, Emily Zakem, et al. "Organic agriculture and the global food supply." *Renewable Agriculture and Food Systems* 22, no. 2 (June 9, 2006): 86–108. Published online July 4, 2007. https://doi.org/10.1017/S1742170507001640.

Balish, Chris. How to *Live Well Without Owning a Car: Save Money, Breathe Easier, and Get More Mileage Out of Life.* Ten Speed Press, 2006.

Barber, Benjamin. *Consumed: how markets corrupt children, infantilize adults, and swallow citizens whole.* W. W. Norton, 2008.

Bardi, Ugo. Extracted: How the Quest for Mineral Wealth is Plundering the Planet. Chelsea Green Publishing, 2014.

---. "Extracting Minerals from Seawater: An Energy Analysis," *Sustainability* 2 (April 9, 2010): 980–992. https://doi.org/10.3390/su2040980.

---. *The Limits to Growth Revisited.* Springer, 2011.

---. "Peak oil, 20 years later: Failed prediction or useful insight?" *Energy Research & Social Science* 48 (2019), 257-261, ISSN 2214-6296, https://doi.org/10.1016/j.erss.2018.09.022.

---. *The Seneca Effect: Why Growth is Slow but Collapse is Rapid.* Springer, 2017.

Barlow, Nathan, Livia Regen, Noémie Cadiou, Ekaterina Chertkovskaya, Max Hollweg, Christina Plank, Merle Schulken and Verena Wolf, editors. *Degrowth and Strategy: How to Bring About Social-Ecological Transformation* (MayFly Books, 2022). https://mayflybooks.org/.

Barlow, Nora, editor. *The Autobiography of Charles Darwin.* W. W. Norton and Company, 1958.

Barnard, Neal D., Joshua Cohen, David J. A. Jenkins, Gabrielle Turner-McGrievy, et al. "A low-fat vegan diet improves glycemic control and cardiovascular risk factors in a randomized clinical trial in individuals with type 2 diabetes." *Diabetes Care* 29, no. 8 (August 2006): 1777–1783. https://doi.org/10.2337/dc06-0606.

Barnes, B. Davis, Judith A. Sclafani, Andrew Zaffos. "Dead clades walking are a pervasive macroevolutionary pattern." *Proceedings of the National Academy of Sciences*, 118, no. 15 (2021), e2019208118, https://doi.org/10.1073/pnas.2019208118.

Barnosky, Anthony. *Dodging Extinction.* University of California Press, 2014.

---. "Megafauna biomass tradeoff as a driver of Quaternary and future extinctions." *Proceedings of the National Academy of Sciences*, 105, Supplement 1 (August 2008): 11543–11548. https://doi.org/10.1073/pnas.0801918105.

Bar-On, Yinon M., Rob Phillips, and Ron Milo. "The biomass distribution on Earth." *Proceedings of the National Academy of Sciences* 115, no. 25 (June 2018): 6506–6511. https://doi.org/10.1073/pnas.1711842115.

Bastin, Jean-Francois, Yelena Finegold, Claude Garcia, Danilo Mollicone, et al. "The global tree restoration potential." *Science* 365, no. 6448 (July 5, 2019): 76–79. https://doi.org/10.1126/science.aax0848.

Batkins, Sam, Philip Rossetti, and Dan Goldbeck. "Putting Nuclear Regulatory Costs in Context." American Action Forum (website), July 12, 2017. https://www.americanactionforum.org/research/putting-nuclear-regulatory-costs-context/.

Bay Area Lyme Foundation (website). "Lyme Disease Facts and Statistics," n.d. (2019?), accessed April 3, 2020. https://www.bayarealyme.org/about-lyme/lyme-disease-facts-statistics/.

Berkes, Howard. "As Mine Protections Fail, Black Lung Cases Surge." National Public Radio, July 9, 2012. https://www.npr.org/2012/07/09/155978300/as-mine-protections-fail-black-lung-cases-surge.

Berwyn, Bob. "Massive Permafrost Thaw Documented in Canada, Portends Huge Carbon Release." *InsideClimate News*, February 28, 2017. https://insideclimatenews.org/news/27022017/global-warming-permafrost-study-melt-canada-siberia.

Bishop, P. J., A. Angulo, J. P. Lewis, R. D. Moore, G. B. Rabb, and J. Garcia Moreno. "The Amphibian Extinction Crisis – what will it take to put the action into the Amphibian Conservation Action Plan?" *S.A.P.I.EN.S* [online], August 12, 2012. URL: http://journals.openedition.org/sapiens/1406.

Bizarro, Sara. "United States: Andrew Yang is running for President in 2020 on the platform of Universal Basic Income." Basic Income Earth Network (website), April 8, 2018. https://basicincome.org/news/2018/04/united-states-andrew-yang-is-running-for-president-in-2020-on-the-platform-of-universal-basic-income/.

Bonometti, Joe. "The Liquid Fluoride Thorium Reactor: What Fusion Wanted to Be." Google TechTalks, November 18, 2008. YouTube video of lecture, 55:16, https://www.youtube.com/watch?v=AHs2Ugxo7-8&feature=youtu.be.

Borlaug, Norman. "Nobel Lecture." The Nobel Prize (website), December 11, 1970. https://www.nobelprize.org/prizes/peace/1970/borlaug/lecture/.

Brambila, Nicole C. "Ogallala Aquifer's dramatic drying sows deep concerns for High Plains agriculture." *Topeka Capital-Journal*, August 12, 2014. https://www.cjonline.com/story/news/politics/state/2014/08/12/ogallala-aquifers-dramatic-drying-sows-deep-concerns-high-plains-agriculture/16660396007/.

Bregman, Rutger. *Utopia for Realists: How We Can Build the Ideal World*. Little, Brown and Co., 2017.

Bricker, Darrell and John Ibbitson. *Empty Planet: The Shock of Global Population Decline*. Crown Publishing Group, 2019.

Broad, William J., and Sergio Peçanha. "The Iran Nuclear Deal – A Simple Guide." *New York Times*, January 15, 2015. https://www.nytimes.com/interactive/2015/03/31/world/middleeast/simple-guide-nuclear-talks-iran-us.html.

Brown, Casey, Julia Mathew, Ilana Wolf, and Ashley Kerkhoff. "What Does 'Plant-based' Actually Mean?" *Vegetarian Journal* 37, no. 4, 2018, 24–28.

Burdon, Peter. "Population is not the problem." ABC Environment (Australia), January 22, 2014. Accessed September 12, 2020. https://web.archive.org/web/20150420042007/www.abc.net.au/environment/articles/2014/01/22/3926001.htm.

Butzer, K. W. "Accelerated Soil Erosion: A Problem of Man-Land Relationships." In *Perspectives on Environment*, ed. I. R. Manners and M. W. Mikesell. Washington, D. C.: Association of American Geographers, 1974.

Campbell, Colin, and Jean Laherrère. "The End of Cheap Oil." *Scientific American*, March 1998, 78–83.

Campbell, T. Colin, and Thomas M. Campbell. *The China Study: The Most Comprehensive Study of Nutrition Ever Conducted and the Startling Implications for Diet, Weight Loss, and Long-term Health.* Ben Bella Books, 2006.

Carlisle, Liz. Lentil Underground: Renegade Farmers and the Future of Food in America. Gotham Books, 2015.

Carrington, Damian. "Plummeting insect numbers 'threaten collapse of nature.'" *The Guardian*, February 10, 2019. https://www.theguardian.com/environment/2019/feb/10/plummeting-insect-numbers-threaten-collapse-of-nature.

Case Western Reserve University. "Gorillas go green: Apes shed pounds while doubling calories on leafy diet, researcher finds." *ScienceDaily*, February 21, 2011. https://www.sciencedaily.com/releases/2011/02/110217091130.htm.

CBC News. "'The anger is getting louder.'" October 24, 2019. https://www.cbc.ca/news/canada/london/vegan-anger-london-verbal-assault-1.5332373.

Centers for Disease Control and Prevention. "About HIV." July 14, 2020. https://www.cdc.gov/hiv/basics/whatishiv.html.

---. *Antibiotic Resistance Threats in the United States, 2019.* Atlanta, GA: U.S. Department of Health and Human Services, CDC. December 2019. https://www.cdc.gov/drugresistance/pdf/threats-report/2019-ar-threats-report-508.pdf.

---. "Information about Middle East Respiratory Syndrome (MERS)." December 2015. https://www.cdc.gov/coronavirus/mers/downloads/factsheet-mers_en.pdf.

---. "Lyme Disease Charts and Figures: Historical Data." November 22, 2019. www.cdc.gov/lyme/stats/graphs.html.

---. "1918 Pandemic (H1N1 virus)." Reviewed March 20, 2019. https://www.cdc.gov/flu/pandemic-resources/1918-pandemic-h1n1.html.

---. "Rocky Mountain Spotted Fever (RMSF)." April 7, 2020. https://www.cdc.gov/rmsf/stats/index.html.

---. "Severe Acute Respiratory Syndrome. Fact Sheet: Basic Information about SARS." January 13, 2004. https://www.cdc.gov/sars/about/fs-SARS.pdf.

Chan, Jasper Fuk-Woo, Kelvin Kai-Wang To, Herman Tse, Dong-Yan Jin, and Kwok-Yung Yuen. "Interspecies transmission and emergence of novel viruses: lessons from bats and birds." *Trends in Microbiology* 21, no. 10 (October 2013), 544–55. https://doi.org/10.1016/j.tim.2013.05.005.

Chandler, David. "Leaving our mark." *MIT News*, April 16, 2008. https://news.mit.edu/2008/footprint-tt0416.

Cho, Renee. "Rare Earth Metals: Will We Have Enough?" Earth Institute, Columbia University. September 19, 2012. https://blogs.ei.columbia.edu/2012/09/19/rare-earth-metals-will-we-have-enough/.

Chung, Emily. "Most groundwater is effectively a non-renewable resource, study finds." CBC News, November 16, 2015. https://www.cbc.ca/news/technology/groundwater-study-1.3318137.

Cichocki, Mark, RN. "How Many People Have Died of HIV?" VeryWell Health (website). This article is periodically updated and was last accessed June 9, 2022. https://www.verywellhealth.com/how-many-people-have-died-of-aids-48721.

Cigainero, Jake. "Who Are France's Yellow Vest Protesters, And What Do They Want?" NPR. December 3, 2018. https://www.npr.org/2018/12/03/672862353/who-are-frances-yellow-vest-protesters-and-what-do-they-want.

Clack, Christopher T. M., Staffan A. Qvist, Jay Apt, Morgan Bazilian, et al. "Evaluation of a proposal for reliable low-cost grid power with 100% wind, water, and solar." *Proceedings of the National Academy of Sciences* 114, no. 26 (June 2017): 6722–6727. https://doi.org/10.1073/pnas.1610381114.

Coaston, Jane. "Try to Resist the Call of the Doomers." *New York Times*, July 23, 2022. https://www.nytimes.com/2022/07/23/opinion/climate-doomers-possibility.html

Cobb, Kurt. "Energy: The Achilles Heel of the Resource Pyramid." *Scitizen* (blog), May 22, 2009. https://www.scitizen.com/future-energies/energy-the-achilles-heel-of-the-resource-pyramid_a-14-2760.html.

---. "How Many Windmills Does It Take to Power the World?" *Scitizen* (blog), February 26, 2008. https://www.scitizen.com/future-energies/how-many-windmills-does-it-take-to-power-the-world-_a-14-1487.html.

Cohen, Joel E. *How Many People Can the Earth Support?* W. W. Norton, 1995.

Cohen, Mark Nathan. *The Food Crisis in Prehistory: Overpopulation and the Origins of Agriculture.* Yale University Press, 1977.

Connor, D. J. "Organic agriculture cannot feed the world." *Field Crops Research* 106, no. 2 (2008): 187–190. https://doi.org/10.1016/j.fcr.2007.11.010.

Coote, Anna, and Andrew Percy. *The Case for Universal Basic Services.* Polity Press, 2020.

Cornell Chronicle. "U.S. could feed 800 million people with grain that livestock eat, Cornell ecologist advises animal scientists." August 7, 1997. https://news.cornell.edu/stories/1997/08/us-could-feed-800-million-people-grain-livestock-eat.

Daly, Herman. *Beyond Growth: The Economics of Sustainable Development.* Beacon Press, 1996.

---."From a Failed Growth Economy to a Steady-State Economy." United States Society for Ecological Economics bi-annual meeting at American University, Washington, D.C., June 1, 2009. https://editors.eol.org/eoearth/wiki/From_a_Failed_Growth_Economy_to_a_Steady-State_Economy.

---. "A Population Perspective on the Steady State Economy." January 15, 2015. Center for the Advancement of the Steady-State Economy. https://steadystate.org/a-population-perspective-on-the-steady-state-economy/.

---. "Toward Some Operational Principles of Sustainable Development." *Ecological Economics* 2, no. 1 (April 1990): 1–6. https://doi.org/10.1016/0921-8009(90)90010-R.

Daly, Herman E., and Joshua Farley, *Ecological Economics: Principles and Applications*. Island Press, 2004.

Daly, Lew, and Stephen Posner. "Beyond GDP: New Measures for a New Economy." Demos, 2011. https://www.demos.org/sites/default/files/publications/BeyondGDP_0.pdf.

Darwin, Charles. *On the Origin of Species*. London, 1859. Penguin Books, 2009.

Dass, Pawlok, Benjamin Z. Houlton, Yingping Wang, and David Warlind. "Grasslands may be more reliable carbon sinks than forests in California." *Environ Res Lett* 13, no. 7 (July 2018): 074027. https://doi.org/10.1088/1748-9326/aacb39.

Davis, Brenda, and Vesanto Melina, *Becoming Vegan. Comprehensive Edition. The Complete Reference to Plant-Based Nutrition*. Book Publishing Company, 2014.

De Bell, Garrett, editor. *The Environmental Handbook*. Ballantine Books, 1970.

de la Croix, David, and Axel Gosseries. "Population Policy through Tradable Procreation Entitlements," CORE Discussion Paper 2006/81, December 5, 2006. https://doi.org/10.2139/ssrn.949148.

De Vos, Jurriaan M., Lucas N. Joppa, John L. Gittleman, Patrick R. Stephens, and Stuart L. Pimm. "Estimating the normal background rate of species extinction." *Conservation Biology*, 29 (April 2015): 452–462. https://doi.org/10.1111/cobi.12380.

Deffeyes, Kenneth S. *Beyond Oil: The View from Hubbert's Peak*. New York: Hill and Wang, 2005.

Diamond, Jared. *Guns, Germs, and Steel: The Fates of Human Societies*. W. W. Norton, 1997.

Dittmar, Michael. "The End of Cheap Uranium." In Bardi, *Extracted*, 64–68.

Donlan, Josh. "Re-wilding North America." *Nature* 436, no. 18 (August 2005): 913–914. https://doi.org/10.1038/436913a.

Drennen, Ari, and Sally Hardin. "Climate Deniers in the 117th Congress." Center for American Progress (website), March 30, 2021. https://www.americanprogress.org/article/climate-deniers-117th-congress/.

Drollette Jr, Dan. "Fusion's greatest hits, as detailed in the Bulletin," *Bulletin of the Atomic Scientists*, April 30, 2019. https://thebulletin.org/2019/04/fusions-greatest-hits-as-detailed-in-the-bulletin/.

Earthworks Group. *50 Simple Things You Can Do to Save The Earth.* Earthworks Press, 1989, 1995.

Earthlings (film). Directed by Shaun Monson, narrated by Joaquin Phoenix. First released

Ehrlich, Paul. *The Population Bomb.* Ballantine Books, 1968. Note: Anne Ehrlich, his wife, was an uncredited co-author of this book.

---. "The Population Crisis: Where We Stand," in Noël Hinrichs, ed., *Population, Environment, and People* (McGraw Hill, 1971), 8–16.

Ehrlich, Paul, and Anne Ehrlich. "Can a Collapse of Civilization Be Avoided?" Simplicity Institute Report 13a. Simplicity Institute, 2013. http://simplicityinstitute.org/wp-content/uploads/2011/04/EhrlichPaper.pdf.

Eisnitz, Gail. *Slaughterhouse: The Shocking Story of Greed, Neglect, and Inhumane Treatment Inside the U.S. Meat Industry.* Prometheus Books, 2006.

Elhacham, Emily, Liad Ben Uri, Jonathan Grozovski, Yinon M. Bar-On, and Ron Milo. "Global human-made mass exceeds all living biomass." *Nature* 588 (October 9, 2020): 442–444. https://doi.org/10.1038/s41586-020-3010-5.

Erb, Karl-Heinz, Thomas Kastner, Christoph Plutzar, Anna Liza S. Bais, et al. "Unexpectedly large impact of forest management and grazing on global vegetation biomass." *Nature* 553 (December 20, 2017): 73–76. https://doi.org/10.1038/nature25138.

Esselstyn, Jr., Caldwell. *Prevent and Reverse Heart Disease.* Avery, 2007.

Expert Panel on Wind Turbine Noise and Human Health. "Understanding the Evidence: Wind Turbine Noise." Council of Canadian Academies. April 9, 2015. https://cca-reports.ca/reports/understanding-the-evidence-wind-turbine-noise/.

Fair, Helen, and Roy Walmsley. *World Prison Population List,* 13th edition. Institute for Crime and Justice Policy Research, December 1, 2021. https://www.prisonstudies.org/sites/default/files/resources/downloads/world_prison_population_list_13th_edition.pdf.

Foer, Jonathan Safran. *Eating Animals.* Little, Brown and Co., 2009.

---. *We Are the Weather: Saving the Planet Begins at Breakfast.* Farrar, Straus, and Giroux, 2019.

Fonseca, Maria João, Nuno H. Franco, Francis Brosseron, Fernando Tavares, I. Anna S. Olsson, and Júlio Borlido-Santos. "Children's attitudes towards animals: evidence from the RODENTIA project." *Journal of Biological Education* Vol. 45 (3), 2011. https://doi.org/10.1080/00219266.2011.576259.

Food and Agriculture Organization of the United Nations. "Forests and climate change: carbon and the greenhouse effect." Forests and Climate Change Working Paper 1, n.d. (2001?), accessed August 19, 2019. http://www.fao.org/3/ac836e/AC836E03.htm.

---. "What are the environmental benefits of organic agriculture?" Organic Agriculture (website), February 2014. http://www.fao.org/organicag/oa-faq/oa-faq6/en/.

Foster, John Bellamy. *The Ecological Revolution: Making Peace with the Planet.* Monthly Review Press, 2009.

Franco, Lara S., Danielle F. Shanahan, and Richard A. Fuller, R. A. (2017). "A Review of the Benefits of Nature Experiences: More Than Meets the Eye." *International Journal of Environmental Research and Public Health* 14(8) (August 2017), 864. https://doi.org/10.3390/ijerph14080864.

Freud, Sigmund. *Totem and Taboo.* W. W. Norton & Company, 1990.

Friedemann, Alice. "Diesel is finite. Trucks are the bedrock of civilization. So where are the battery electric trucks?" *EnergySkeptic.com* (blog), August 17, 2021. https://energyskeptic.com/2021/diesel-finite-where-are-electric-trucks/.

---. *Life After Fossil Fuels: A Reality Check on Alternative Energy.* Lecture Notes in Energy, volume 81. Springer, 2021.

Friedmann, S. Julio, Zhiyuan Fan, and Ke Tang. "Low-carbon Heat Solutions for Heavy Industry: Sources, Options, and Costs Today." Center on Global Energy Policy, Columbia School of International and Political Affairs. October 2019. https://www.energypolicy.columbia.edu/research/report/low-carbon-heat-solutions-heavy-industry-sources-options-and-costs-today.

Fritz, Charles. *Disasters and Mental Health: Therapeutic Principles Drawn from Disaster Studies.* University of Delaware Disaster Research Center, Historical and Comparative Disaster Series #10, 1996. https://udspace.udel.edu/handle/19716/1325.

Fuentes, Augustin. "Bad to the Bone: Are Humans Naturally Aggressive?" *Psychology Today*, April 18, 2012. https://www.psychologytoday.com/blog/busting-myths-about-human-nature/201204/bad-the-bone-are-humans-naturally-aggressive.

Gee, Henry. *A (Very) Short History of Life on Earth: 4.6 Billion Years in 12 Pithy Chapters.* St. Martin's Press, 2021.

Gee, Henry. "Humans Are Doomed to Go Extinct." *Scientific American*, November 30, 2021. https://www.scientificamerican.com/article/humans-are-doomed-to-go-extinct/.

Glaser, Christine, Chuck Romaniello, and Karyn Moskowitz. *Costs and Consequences: The Real Costs of Livestock Grazing on America's Public Lands.* Center for Biological Diversity, 2015. https://www.biologicaldiversity.org/programs/public_lands/grazing/pdfs/CostsAndConsequences_01-2015.pdf.

Gleeson, Tom, Kevin M. Befus, Scott Jasechko, Elco Luijendijk, and M. Bayani Cardenas. "The global volume and distribution of modern groundwater." *Nature Geoscience* 9 (November 16, 2015): 161–167. https://doi.org/10.1038/ngeo2590.

Global Climate Change (website). "Scientific Consensus: Earth's Climate Is Warming." Science Mission Directorate, National Aeronautics and Space Administration (NASA). https://climate.nasa.gov/scientific-consensus/, accessed April 18, 2022.

Glowacki, Luke. "Are people naturally violent? Probably." *Los Angeles Times*, January 19, 2014. http://articles.latimes.com/2014/jan/19/opinion/la-oe-glowacki-violence-humans-genes-20140119.

Goodland, Robert. "FAO Yields to Meat Industry Pressure on Climate Change." *New York Times*, July 11, 2012. https://bittman.blogs.nytimes.com/2012/07/11/fao-yields-to-meat-industry-pressure-on-climate-change/.

---. "How to Reverse Climate Change Before It's Too Late." Chomping Climate Change (website), July 1, 2016. http://www.chompingclimatechange.org/updated-analysis/how-to-reverse-climate-change-before-its-too-late/.

---. "'Livestock and Climate Change': Critical Comments and Responses." *WorldWatch* 23, no. 2 (March 2010): 7–9. https://awfw.org/wp-content/uploads/WWMLivestock-ClimateResponses-20101.pdf.

Goodland, Robert, and Jeff Anhang. "Livestock and Climate Change." *WorldWatch* 22, no. 6 (November-December 2009): 10–19. https://awellfedworld.org/wp-content/uploads/Livestock-Climate-Change-Anhang-Goodland.pdf.

---. "Livestock and greenhouse gas emissions: The importance of getting the numbers right, by Herrero et al." *Anim. Feed Sci. Technol.* 172, no. 3–4 (March 30, 2012): 252–256. https://doi.org/10.1016/j.anifeedsci.2011.12.028. Content accessed at https://awfw.org/wp-content/uploads/pdf/Goodland-Anhang-Livestock-GHG-1-7-12.pdf.

Gore, Al. *Our Choice: A Plan to End the Climate Crisis*. Rodale Press, 2009.

Gorillas-World (website). "Gorilla Feeding." February 6, 2014. https://www.gorillas-world.com/gorilla-feeding/.

Gradidge, Sarah, Magdalena Zawisza, Annelie J. Harvey, and Daragh T. McDermott. "A Structured Literature Review of the Meat Paradox." *Social Psychological Bulletin* 16, no. 3 (September 23, 2021), 1–26. https://doi.org/10.32872/spb.5953.

Grandel, Leena, and Mikael Höök. "Assessing Rare Metal Availability Challenges for Solar Energy Technologies." *Sustainability* 7, no. 9 (August 26, 2015): 11818-11837. https://doi.org/10.3390/su70911818.

Greer, John Michael. "A Terrible Ambivalence." Resilience (website), September 10, 2009. https://www.resilience.org/stories/2009-09-10/terrible-ambivalence/.

Greger, Michael. *Bird Flu: A Virus of Our Own Hatching.* Lantern Books, 2006.

---. *How To Survive a Pandemic.* Flatiron Books, 2020.

---. "The McGovern Report." NutritionFacts.Org (website), April 12, 2013. https://nutritionfacts.org/video/the-mcgovern-report/.

Greger, Michael with Gene Stone. *How Not to Die. Discover the Foods Scientifically Proven to Prevent and Reverse Disease.* Flatiron Books, 2015.

Grossman, Dave. On Killing. *The psychological cost of learning to kill in war and society.* Little, Brown and Co., 2009.

Haberl, Helmut, Karl-Heinz Erb, and Fridolin Krausmann. "Global human appropriation of net primary production (HANPP)." The Encyclopedia of Earth (website), September 3, 2013. https://editors.eol.org/eoearth/wiki/Global_human_appropriation_of_net_primary_production_(HANPP).

Hahn, Erin R., Meghan Gillogly, and Bailey E. Bradford. "Children are unsuspecting meat eaters: An opportunity to address climate change." *Journal of Environmental Psychology* 78 (2021): 10175. https://doi.org/10.1016/j.jenvp.2021.101705.

Hail, Steven. "Paying for a Green New Deal: An Introduction to Modern Monetary Theory." In Williams and Taylor, eds., *Sustainability and the New Economics*, pp. 279–302.

Hamilton, James. "Historical Oil Shocks." Prepared for chapter 21 of the *Routledge Handbook of Major Events in Economic History*, Randall E. Parker and Robert Whaples, editors (Routledge, 2013). https://econweb.ucsd.edu/~jhamilto/oil_history.pdf.

Hansen, James. *Storms Of My Grandchildren: The Truth About the Climate Catastrophe and Our Last Chance to Save Humanity.* Bloomsbury Publishing PLC, 2009.

Harari, Yuval Noah. *Sapiens: A Brief History of Humankind.* Harvill Secker, 2014.

Hardin, Garrett. "The Tragedy of the Commons." In De Bell, Garrett, ed., *The Environmental Handbook.* Ballantine Books, 1970.

Harper, Kristin, and George Armelagos, "The Changing Disease-Scape in the Third Epidemiological Transition." *Int. J. Environ. Res. Public Health* 7, no. 2 (February 24, 2010): 675–697. https://doi.org/10.3390/ijerph7020675.

Hawken, Paul, editor. *Drawdown: The Most Comprehensive Plan Ever Proposed to Reverse Global Warming.* Penguin, 2017.

Hayek, Matthew N., Helen Harwatt, William J. Ripple, and Nathaniel D. Mueller. "The carbon opportunity cost of animal-sourced food production on land." *Nature Sustainability* 4 (2021), pp. 21–24. https://doi.org/10.1038/s41893-020-00603-4.

Heinberg, Richard. *Blackout: Coal, Climate, and the Last Energy Crisis.* New Society Publishers, 2009.

Herman, J. "Saving U.S. Dietary Advice from Conflicts of Interest." *Food and Drug Law Journal* 65, no. 2 (2010), pp. 285–316. https://www.drmcdougall.com/misc/2010nl/jul/sc%20herman.indd.pdf.

Herrero, M., P. Gerber, T. Vellinga, T. Garnett, et al. "Livestock and greenhouse gas emissions: The importance of getting the numbers right," *Anim. Feed Sci. Technol.* 166–167 (June 23, 2011): 779–782. https://doi.org/10.1016/j.anifeedsci.2011.04.083.

Hickel, Jason. *Less is More: How Degrowth Will Save the World.* Windmill Books, 2020.

Hickel, Jason, and Giorgos Kallis. "Is Green Growth Possible?" *New Political Economy* 25, no. 4 (April 17, 2019): 469–486. https://doi.org/10.1080/13563467.2019.1598964.

Hirsch, Robert L., Roger Bezdek, and Robert Wendling. "Peaking of world oil production: Impacts, mitigation, & risk management." February 2005. U.S. Department of Energy, Office of Scientific and Technical Information. https://doi.org/10.2172/939271.

Hodson, Gordon. "Why Vegans Make Some People So Uncomfortable and Angry." *Psychology Today*, October 27, 2019. https://www.psychologytoday.com/intl/blog/without-prejudice/201910/why-vegans-make-some-people-so-uncomfortable-and-angry.

Holdren, John. "A Brief History of 'IPAT.'" *The Journal of Population and Sustainability* 2, no. 2 (Spring 2018): 66–74. https://jpopsus.org/full_articles/holdren-vol2-no2/.

Huang Chaolin, Yeming Wang, Xingwang Li, Lili Ren, et al. "Clinical features of patients infected with 2019 novel coronavirus in Wuhan, China." *Lancet* 395, no. 10223 (February 15, 2020): 497–506. https://doi.org/10.1016/S0140-6736(20)30183-5.

Hubbert, M. King. "Nuclear Energy and the Fossil Fuels." Shell Development Company, Publication No. 95, June 1956.

Humanity Forward (website). https://movehumanityforward.com/.

Humes, Edward. *Door to Door: The Magnificent, Maddening, Mysterious World of Transportation.* Harper Collins, 2016.

Hunt, Melissa G., Rachel Marx, Courtney Lipson, and Jordyn Young. "No More FOMO: Limiting Social Media Decreases Loneliness and Depression." *Journal of Social and Clinical Psychology* 37, no. 10 (2018), pp. 751–768.

Inman, Mason. "How to Measure the True Cost of Fossil Fuels." *Scientific American*, April 1, 2013, 58–61. http://literacy473.weebly.com/uploads/9/1/6/7/9167715/inman_2013_true_cost_of_fossil_fuels_scientificamerican0413-58.pdf.

---. *The Oracle of Oil: A Maverick Geologist's Quest for a Sustainable Future*. W. W. Norton & Company, 2016.

International Energy Agency. *Perspectives on the Energy Transition*. IEA Publications and IRENA Publications, 2017. https://www.irena.org/-/media/Files/IRENA/Agency/Publication/2017/Mar/Perspectives_for_the_Energy_Transition_2017.pdf.

Ivanova, Diana and Milena Büchs. "Implications of shrinking household sizes for meeting the 1.5 °C climate targets." *Ecological Economics* 202 (2022), 107590, ISSN 0921-8009, https://doi.org/10.1016/j.ecolecon.2022.107590.

Jaber, Nadia. "Feeding the Planet, An Evening with Alan Weisman." The New York Academy of Sciences (website), Academy E-briefing, October 31, 2013. Accessed August 25, 2020. https://www.nyas.org/ebriefings/feeding-the-planet/.

Jacobson, Mark Z., and Mark A. Delucchi, "A Plan to Power 100 Percent of the Planet with Renewables," *Scientific American*, November 2009, 58–65. http://web.stanford.edu/group/efmh/jacobson/Articles/I/sad1109Jaco5p.indd.pdf.

Jaspers, Karl. *The origin and goal of history*. Translated by Michael Bullock. Yale University Press, 1953.

Ji, Sayer. "'Million Cancer Deaths from Fukushima Expected in Japan,' New Report Reveals." GreenMedinfo, November 8, 2015. https://www.greenmedinfo.com/blog/million-cancer-deaths-fukushima-expected-japan-new-report-reveals-1.

Johnson, George. "When radiation isn't the real risk." *New York Times*, September 21, 2015. https://www.nytimes.com/2015/09/22/science/when-radiation-isnt-the-real-risk.html.

Johnson, Leonard C. "Soil loss tolerance: fact or myth?" *Journal of Soil and Water Conservation* 42, no. 3 (May–June 1987): 155–160.

Jordan, Douglas. "The Deadliest Flu." December 17, 2019. Centers for Disease Control and Prevention (website). https://www.cdc.gov/flu/pandemic-resources/reconstruction-1918-virus.html.

Jones, Lucy. *Losing Eden: Our Fundamental Need for the Natural World and Its Ability to Heal Body and Soul.* Pantheon Books, 2020, 2021.

Jones, Nicola. "A Scarcity of Rare Metals Is Hindering Green Technologies." Yale Environment 360 (website), November 18, 2013. https://e360.yale.edu/features/a_scarcity_of_rare_metals_is_hindering_green_technologies.

Joy, Melanie. *Why We Love Dogs, Eat Pigs, and Wear Cows.* Conari Press, 2010.

Junger, Sebastian. *Tribe: On Homecoming and Belonging.* Twelve, 2016.

Karesh, William B., Robert Cook, Elizabeth L. Bennett, and James Newcomb. "Wildlife Trade and Global Disease Emergence. Emerging Infectious Diseases." *Emerging Infectious Diseases* 11, no. 7 (July 2005): 1000–1002. https://doi.org/10.3201/eid1107.050194.

Karlin, Mark. "Animal Slaughter Industry Making It Illegal to Show You Cruelty, Including Veal Calves Skinned Alive," Buzzflash at Truthout, March 20, 2013. http://truth-out.org/buzzflash/commentary/item/17872-animal-slaughter-industry-making-it-illegal-to-show-you-cruelty-including-veal-calves-skinned-alive.

Kbh3rd at Wikipedia. "High plains fresh groundwater usage." https://en.wikipedia.org/wiki/File:High_plains_fresh_groundwater_usage_2000.svg, accessed September 23, 2020.

Kean, Sam. "Peak Phosphorus?" *Science History Institute*, January 5, 2014. https://www.sciencehistory.org/distillations/peak-phosphorus.

Keister, Lisa. "The One Percent," *Annu. Rev. Sociol.* 40 (April 3, 2014): 347–367. https://doi.org/10.1146/annurev-soc-070513-075314. Text online at: https://wealthinequality.org/wp-content/uploads/2015/10/Keister-The-One-Percent.pdf.

Kharecha, Pushker A., and James E. Hansen. "Prevented Mortality and Greenhouse Gas Emissions from Historical and Projected Nuclear Power." *Environ. Sci. Technol.* 47 (2013): 4889–4895. https://dx.doi.org/10.1021/es3051197.

Kilbourne, Jean. *Can't buy my love: how advertising changes the way we think and feel.* Simon & Schuster, 2000.

Kimani, Alex. "The Holy Grail of Energy is Finally Within Reach." *OilPrice.com*, October 21, 2019. https://oilprice.com/Energy/ Energy-General/The-Holy-Grail-Of-Energy-Is-Finally-Within-Reach.html.

Klein, Naomi. *This Changes Everything: Capitalism vs. the Climate.* Simon and Schuster, 2014.

Koch, Alexander, Chris Brierley, Mark M. Maslin, and Simon L. Lewis. "Earth system impacts of the European arrival and Great Dying in the Americas after 1492." *Quaternary Science Reviews* 207 (March 1, 2019): 13–36. https://doi.org//10.1016/j. quascirev.2018.12.004.

Koch, Paul, and Anthony Barnosky. "Late Quaternary Extinctions: State of the Debate," *Annual Review of Ecology, Evolution, and Systematics* 37 (August 25, 2006): 215–250. https://doi. org/10.1146/annurev.ecolsys.34.011802.132415.

Kohn, Alfie. "Are Humans Innately Aggressive?" Originally published in *Psychology Today*, June 1988. https://www. alfiekohn.org/article/humans-innately-aggressive/.

Kokelj, Steven V., Trevor C. Lantz, Jon Tunnicliffe, Rebecca Segal, and Denis Lacelle. "Climate-driven thaw of permafrost preserved glacial landscapes, northwestern Canada." *Geology* 45, no. 4 (April 1, 2017): 371–374. https://doi.org/10.1130/ G38626.1.

Koppitke, Peter, Neal W. Menzies, Peng Wang, Brigid A. McKenna, and Enzo Lombi. "Soil and the intensification of agriculture for global food security." *Environment International* 132 (2019), 105078. https://doi.org/10.1016/j.envint.2019.105078.

Kovda, V. A. "Soil Reclamation and Food Production." In *Food, Climate, and Man*, ed. M. R. Biswas and A. K. Biswas. New York: John Wiley & Sons, 1979.

Kovel, Joel. *The Enemy of Nature: The End of Capitalism or the End of the World?* Zed Books, 2002.

Krausmann, Fridolin, Karl-Heinz Erb, Simone Gingrich, Helmut Haberl, et al. "Global human appropriation of net primary production doubled in the 20th century." *Proceedings of the National Academy of Sciences* 110, no. 25 (June 2013): 10324-10329. https://doi.org/10.1073/pnas.1211349110.

Kucharik, Christopher J., and Navin Ramankutty. "Trends and Variability in U.S. Corn Yields Over the Twentieth Century." *Earth Interactions* 9, no. 1 (March 2005): 1–29. https://doi.org/10.1175/EI098.1.

Kuhn, Thomas S. *The Structure of Scientific Revolutions.* The University of Chicago Press, 1962.

Kunst, Jonas R., and Sigrid M. Hohle. "Meat eaters by dissociation: How we present, prepare and talk about meat increases willingness to eat meat by reducing empathy and disgust." *Appetite* 105 (2016), 758–774. https://doi.org/10.1016/j.appet.2016.07.009.

Kunstler, James Howard. *The Geography of Nowhere: The Rise and Decline of America's Man-Made Landscape.* Simon and Schuster, 1993.

---. *The Long Emergency: Surviving the End of Oil, Climate Change, and the Other Converging Catastrophes of the Twenty-First Century.* Grove Press, 2005.

Kurtzleben, Danielle. "Rep. Alexandria Ocasio-Cortez Releases Green New Deal Outline." NPR, February 7, 2019. https://www.npr.org/2019/02/07/691997301/rep-alexandria-ocasio-cortez-releases-green-new-deal-outline.

Kwiatkowska, Teresa. "The Sadness of the woods is bright: Deforestation and conservation in the middle ages," *Medievalia* 39 (January 2007): 40–47.

Lang, Olivia. "The dangers of mining around the world." BBC News, October 14, 2010. https://www.bbc.com/news/world-latin-america-11533349.

Lappé, Frances Moore. *Diet for a Small Planet.* Revised edition. Ballantine Books, 1975.

Laurence, Charles. "US farmers fear the return of the Dust Bowl." *The Telegraph*, March 7, 2011. https://www.telegraph.co.uk/news/earth/8359076/US-farmers-fear-the-return-of-the-Dust-Bowl.html.

Laville, Sandra. "Dumped fishing gear is biggest plastic polluter in ocean, finds report," *The Guardian*, November 6, 2019. https://www.theguardian.com/environment/2019/nov/06/dumped-fishing-gear-is-biggest-plastic-polluter-in-ocean-finds-report.

Lawrence, Kate. *The Practical Peacemaker: How Simple Living Makes Peace Possible.* Lantern Books, 2009.

LeBlanc, Steven, and Katherine E. Register. *Constant Battles: The Myth of the Peaceful, Noble Savage.* St. Martin's Press, 2003.

Leenaert, Tobias. *How to Create a Vegan World: A Pragmatic Approach.* Lantern Books, 2017.

Leite, Ana C., Kristof Dhont, and Gordon Hodson. (2019), "Longitudinal effects of human supremacy beliefs and vegetarianism threat on moral exclusion (vs. inclusion) of animals." *Eur. J. Soc. Psychol.* 49: 179–189. https://doi.org/10.1002/ejsp.2497.

Lelyveld, Michael. "China Cuts Coal Mine Deaths, but Count in Doubt." Radio Free Asia, March 16, 2015. https://www.rfa.org/english/commentaries/energy_watch/china-coal-deaths-03162015103452.html.

Letzter, Rafi. "A massive solar storm could wipe out almost all of our modern technology." Business Insider (website), September 6, 2016. https://www.businessinsider.com/massive-1859-solar-storm-telegraph-scientists-2016-9.

Levin, Diane E., and Jean Kilbourne. *So sexy so soon: the new sexualized childhood, and what parents can do to protect their kids.* Ballantine Books, 2008.

Linkola, Pentti. *Eco-Fascism Writings*, "The Sum of Life," translated September 14, 2006. http://www.penttilinkola.com/pentti_linkola/ecofascism_writings/translations/voisikoelamavoittaa_translation/VII%20-%20The%20Prerequisites%20Of%20Life/.

Lister, Bradford C., and Andres Garcia. "Climate-driven declines in arthropod abundance restructure a rainforest food web." *Proceedings of the National Academy of Sciences*, 115, no. 44 (October 30, 2018): E10397-E10406. https://doi.org/10.1073/pnas.1722477115.

Loomis, Brandon. "Pipelines? Desalination? Turf removal? Arizona commits $1B to augment, conserve water supplies." AZCentral.com (website), June 27, 2022. https://www.azcentral.com/story/news/local/arizona-environment/2022/06/27/arizona-lawmakers-bank-billion-dollars-augment-and-save-water/7736861001/.

Louv, Richard. *Last Child in the Woods: Saving Our Children from Nature-Deficit Disorder* (Algonquin Books of Chapel Hill, 2008).

Lowrey, Annie. *Give People Money: How a Universal Basic Income Would End Poverty, Revolutionize Work, and Remake the World.* Crown, 2018.

Lundholm, B. "Domestic Animals in Arid Ecosystems." In Anders Rapp et al., eds., *Can Desert Encroachment be Stopped?*

MacInnis, Cara C., and Gordon Hodson. "It Ain't Easy Eating Greens: Evidence of Bias toward Vegetarians and Vegans from Both Source and Target." *Group Processes & Intergroup Relations*, 20, no. 6 (Nov. 2017), pp. 721–744. https://doi.org/10.1177/1368430215618253.

MacKay, David J. C. *Sustainable Energy—-without the hot air.* UIT, Cambridge, England, 2009. http://withouthotair.com/cft.pdf.

MacNair, Rachel. *Perpetration-Induced Traumatic Stress: The Psychological Consequences of Killing.* Praeger/Greenwood, 2005.

Mallapaty, Smriti. "China prepares to test thorium–fuelled nuclear reactor." *Nature* 597 (September 16, 2021): 311–312. https://www.nature.com/articles/d41586-021-02459-w.

Malthus, Thomas. *An Essay on the Principle of Population.* London, 1798. The Electronic Scholarly Publishing Project, 1998. http://www.esp.org/books/malthus/population/malthus.pdf.

Mandel, Emily St. John. *Station Eleven.* Knopf, 2014.

Manfreda, John. "The Real History of Fracking." *OilPrice.com*, April 13, 2015. https://oilprice.com/Energy/Crude-Oil/The-Real-History-Of-Fracking.html.

Manno, Jack. *Privileged Goods: Commoditization and Its Impact on Environment and Society.* Lewis Publishers, 2000.

Marchese, David. "This Eminent Scientist Says Climate Activists Need to Get Real." *New York Times*, April 22, 2022. https://www.nytimes.com/interactive/2022/04/25/magazine/vaclav-smil-interview.html.

Marris, Emma. *Wild Souls: Freedom and Flourishing in the Non-Human World.* Bloomsbury Publishing, 2021.

Martin, Paul S. "Prehistoric Overkill: The Global Model." In *Quaternary Extinctions: A Prehistoric Revolution*, Paul S. Martin, Richard G. Klein, editors. The University of Arizona Press, Tucson, Arizona, 1984.

Marx, Karl. *Critique of the Gotha Program*, part I. Widely available online and in anthologies, e.g., Robert Tucker, editor, *The Marx-Engels Reader* (Norton, 1978), and https://www.marxists.org/archive/marx/works/download/Marx_Critque_of_the_Gotha_Programme.pdf.

Matson, P. A., W. J. Parton, A. G. Power, M. J. Swift. "Agricultural Intensification and Ecosystem Properties." *Science* 277, no. 5325 (July 25, 1997): 504–509. https://doi.org/10.1126/science.277.5325.504.

McLaughlin, Michael, Carrie Pettus-Davis, Derek Brown, Chris Veeh, and Tanya Renn. "The Economic Burden of Incarceration in the U.S." Concordance Institute for Advancing Social Justice, Working Paper #CI072016, July 2016. http://joinnia.com/wp-content/uploads/2017/02/The-Economic-Burden-of-Incarceration-in-the-US-2016.pdf.

McMahon, Jeff. "Meat and Agriculture Are Worse for the Climate Than Power Generation, Steven Chu Says." *Forbes*, April 4, 2019. https://www.forbes.com/sites/jeffmcmahon/2019/04/04/meat-and-agriculture-are-worse-for-the-climate-than-dirty-energy-steven-chu-says/.

McMahon, Kate, and Victoria Witting. "Corn ethanol and climate change." Friends of the Earth, July 2011. https://foe.org/wp-content/uploads/2017/legacy/Corn_ethanol_and_climate_change.pdf.

McNeill, J. R. *Something New Under the Sun: An Environmental History of the Twentieth-Century World*. W. W. Norton, 2000.

McWilliams, James. "PTSD in the Slaughterhouse." *Texas Observer*, February 7, 2012. https://www.texasobserver.org/ptsd-in-the-slaughterhouse/.

Meadows, Donella H., Dennis L. Meadows, Jørgen Randers, and William H. Behrens III. *The Limits to Growth: A Report for the Club of Rome's Project on the Predicament of Mankind*. New American Library, 1972.

Mearns, Euan. "ERoEI for beginners." *Energy Matters* (blog), May 25, 2016. http://euanmearns.com/eroei-for-beginners/.

Merrill, Dave, and Lauren Leatherby. "Here's How America Uses Its Land." *Bloomberg*. July 31, 2018. https://www.bloomberg.com/graphics/2018-us-land-use/.

Meyer, François, Isabelle Bairati, Ramak Shadmani, Yves Fradet, and Lynne Moore. "Dietary fat and prostate cancer survival." *Cancer Causes Control* 10 (1999): 245–251. https://doi.org/10.1023/A:1008913307947.

Meyer, William B., and B. L. Turner II. "Human Population Growth and Global Land-Use/Cover Change." *Annu. Rev. Ecol. Syst.* 23 (1992): 39–61. www.jstor.org/stable/2097281.

Milman, Oliver. "Suicides indicate wave of 'doomerism' over escalating climate crisis." The Guardian, May 19, 2022. https://www.theguardian.com/environment/2022/may/19/climate-suicides-despair-global-heating

Modlinska, Klaudia, and Wojciech Pisula. "Selected Psychological Aspects of Meat Consumption—A Short Review." *Nutrients* 10 (9) (September 2018): 1301. https://doi.org/10.3390/nu10091301.

Monbiot, George. "Clearing up this mess." November 18, 2008. *George Monbiot* (blog). https://www.monbiot.com/2008/11/18/clearing-up-this-mess/.

---. "Destroyer of Worlds." *The Guardian*, March 25, 2014. https://www.monbiot.com/2014/03/24/destroyer-of-worlds/.

---. "Evidence Meltdown." *The Guardian*, April 5, 2011. https://www.monbiot.com/2011/04/04/evidence-meltdown/.

---. "We can't keep eating as we are—-why isn't the IPCC shouting this from the rooftops?" *The Guardian*, August 8, 2019. https://www.theguardian.com/commentisfree/2019/aug/08/ipcc-land-climate-report-carbon-cost-meat-dairy.

Montgomery, David R. *Growing a Revolution: Bringing Our Soil Back to Life*. W. W. Norton, 2017.

Moore, Megan. "Basins: Alberta Tar Sands." Fossil Fuel Connections (website), Evergreen State College, 2015–2016. http://www.fossilfuelconnections.org/alberta-tar-sands.

Mosher, Dave. "Gorillas More Related to People Than Thought, Genome Says." *National Geographic News*, March 8, 2012. https://www.nationalgeographic.com/news/2012/3/120306-gorilla-genome-apes-humans-evolution-science/.

Murray, Christopher J. L., and GBD 2017 Population and Fertility Collaborators. "Population and fertility by age and sex for 195 countries and territories, 1950–2017." *Lancet* 392 (November 10, 2018): 1995–2051. https://www.thelancet.com/pdfs/journals/lancet/PIIS0140-6736(18)32278-5.pdf.

National Commission on the BP Deepwater Horizon Oil Spill and Offshore Drilling. "A Brief History of Offshore Oil Drilling," Draft, Staff Working Paper No. 1, n.d. (2010?), accessed March 13, 2017. https://web.cs.ucdavis.edu/~rogaway/classes/188/materials/bp.pdf.

Nhat Hanh, Thich. *Living Buddha, Living Christ*. Riverhead Books, 2007.

---. *Zen and the Art of Saving the Planet*. HarperCollins Publishers, 2021.

Nhat Hanh, Thich, and the Monks and Nuns of Plum Village. *One Buddha is Not Enough: A Story of Collective Awakening*. Parallax Press, 2010.

Nordhaus, William. "Baumol's Diseases: a Macroeconomic Perspective." National Bureau of Economic Research, Working Paper 12218, May 2006. https://doi.org/10.3386/w12218.

---. *The climate casino: risk, uncertainty, and economics for a warming world*. Yale University Press, 2013.

Novikoff, G. "Traditional Grazing Practices and Their Adaptation to Modern Conditions in Tunisia and the Sahelian Countries," in Anders Rapp et al., eds., *Can Desert Encroachment be Stopped?*

Nunez, Christina. "Deforestation explained." http://www.nationalgeographic.com/environment/global-warming/deforestation/, accessed February 23, 2019.

Odum, Howard T., and Elisabeth C. Odum. *A Prosperous Way Down*. University Press of Colorado, 2001.

Office of the Governor Doug Ducey (website). "Governor Ducey Signs Legislation to Secure Arizona's Water Future" (news release), July 6, 2022. https://azgovernor.gov/governor/news/2022/07/governor-ducey-signs-legislation-secure-arizonas-water-future

O'Laughlin, Jay, and Ron Mahoney. "Forests and Carbon." *Woodland NOTES* 19, no. 2 (Spring/Summer 2008). Policy Analysis Group (PAG) Issue Brief No. 11. https://idahoforests.org/wp-content/uploads/forests-and-carbon.pdf.

Oppenheimer, Michael, Naomi Oreskes, Dale Jamieson, Keynyn Brysse, Jessica O'Reilly, Matthew Shindell, and Milena Wazeck. *Discerning Experts: The Practices of Scientific Assessment for Environmental Policy*. University of Chicago Press, 2019.

Orr, David W. "Love it or Lose It: The Coming Biophilia Revolution." In *The Biophilia Hypothesis*, Stephen R. Kellert and Edward O. Wilson, editors (Island Press, 1993), pp. 415–440.

Oswald, James, Mike Raine, and Hezlin Ashraf-Ball. "Will British Weather Provide Reliable Electricity?" *Energy Policy* 36, no. 8 (August 2008): 3212–3225. https://doi.org/10.1016/j.enpol.2008.04.033.

Our Hen House (website). "Interview with Dr. T. Colin Campbell." September 14, 2013. https://www.ourhenhouse.org/episode-192-if-i-look-at-the-mass-i-will-never-act-if-i-look-at-the-one-i-will/.

Oxfam. "Extreme Carbon Inequality." Oxfam Media Briefing, December 2, 2015. https://www.oxfamamerica.org/explore/research-publications/extreme-carbon-inequality/.

Parrique, Timothée, J. Barth, F. Briens, C. Kerschner, A. Kraus-Polk, A. Kuokkanen, and J. H. Spangenberg. "Decoupling debunked: Evidence and arguments against green growth as a sole strategy for sustainability." European Environmental Bureau, July 2019.

Passipedia. "Passipedia – The Passive House Resource." https://passipedia.org/start.

---. "The Passive House – historical review." https://passipedia.org/basics/the_passive_house_-_historical_review.

Patterson, Ron. "Was 2018 the Peak for Crude Oil Production?" *OilPrice.com*, March 14, 2019. https://oilprice.com/Energy/Energy-General/Was-2018-The-Peak-For-Crude-Oil-Production.html.

People for the Ethical Treatment of Animals. "Animal Abuse & Human Abuse: Partners in Crime," n.d. http://www.mediapeta.com/peta/PDF/AnimalAbuseHumanAbuse.pdf, accessed April 3, 2017.

---. "How to Go Vegan." https://how-to-go-vegan.peta.org/.

People's Daily Online. "Black lung disease claims 140,000 lives in China." March 18, 2005. http://en.people.cn/200503/18/eng20050318_177365.html.

Persson, Linn, Bethanie M. Carney Almroth, Christopher D. Collins, Sarah Cornell, et al. "Outside the Safe Operating Space of the Planetary Boundary for Novel Entities." *Environ. Sci. Technol.* 56, no. 3 (2022), 1510–1521. https://doi.org/10.1021/acs.est.1c04158.

Peters, Christian J., Jamie Picardy, Amelia F. Darrouzet-Nardi, Jennifer L. Wilkins, Timothy S. Griffin, and Gary W. Fick. "Carrying capacity of U.S. agricultural land: Ten diet scenarios." *Elementa: Science of the Anthropocene* (2016) 4: 000116. https://doi.org/10.12952/journal.elementa.000116.

Peterson, Per F. "Will the United States Need a Second Geologic Repository?" *The Bridge* 33, no. 3 (Fall 2003), pp. 26–32. https://www.nae.edu/19579/19582/21020/7380/7602/WilltheUnitedStatesNeedaSecondGeologicRepository.

Pew Research Center. "The Chinese Celebrate Their Roaring Economy, As They Struggle With Its Costs." The 2008 Pew Global Attitudes Survey in China. July 22, 2008. https://www.pewresearch.org/global/2008/07/22/the-chinese-celebrate-their-roaring-economy-as-they-struggle-with-its-costs/.

---."For most U.S. workers, real wages have barely budged in decades," by Drew DeSilver. August 7, 2018. https://www.pewresearch.org/fact-tank/2018/08/07/for-most-us-workers-real-wages-have-barely-budged-for-decades/.

Piazza, Jared, Matthew B. Ruby, Steve Loughnan, Mischel Luong, Juliana Kulik, Hanne M. Watkins, and Mirra Seigerman. "Rationalizing meat consumption: The 4Ns." *Appetite* 91 (2015), 114–128. https://doi.org/10.1016/j.appet.2015.04.011.

Pielke, Roger A., J. Adegoke, A. Beltraán-Przekurat, C. A. Hiemstra, et al. "An overview of regional land-use and landcover impacts on rainfall." *Tellus* 59B (2007): 587–601. https://doi.org/10.1111/j.1600-0889.2007.00251.x.

Pimentel, David, Bonnie Berger, David Filberto, Michelle Newton, et al. "Water Resources: Agricultural and Environmental Issues." In Pimentel and Pimentel, *Food, Energy, and Society*, 183–199.

Pimentel, David, and Tad W. Patzek. "Ethanol Production Using Corn, Switchgrass, and Wood; Biodiesel Production Using Soybean and Sunflower." *Nat Resour Res* 14 (2005): 65–76. https://doi.org/10.1007/s11053-005-4679-8.

Pimentel, David, and Marcia Pimentel. *Food, Energy, and Society,* Third edition. CRC Press, 2007.

---. "Soil Erosion: A Food and Environmental Threat." In Pimentel and Pimentel, *Food, Energy, and Society.*

---. "World Population, Food, Natural Resources, and Survival." *World Futures* 59 (2003): 45–167. https://doi.org/10.1080/02604020310124.

Pimentel, David, Michele Whitecraft, Zachary R. Scott, Leixin Zhao, et al. "Will Limited Land, Water, and Energy Control Human Population Numbers in the Future?" *Hum Ecol* 38 (2010): 599–611. https://doi.org/10.1007/s10745-010-9346-y.

Pinker, Stephen. *The Better Angels of Our Nature: Why Violence Has Declined.* Penguin Books, 2011.

Plant Positive (website). "How Time Magazine Sacrificed Its Standards to Promote Saturated Fat." June 28, 2014. http://plantpositive.com/blog/2014/6/28/how-time-magazine-sacrificed-its-standards-to-promote-satura.html.

Plato. *The Collected Dialogues of Plato Including the Letters.* Edith Hamilton and Huntington Cairns, eds. Bollingen Series LXXI. Princeton University Press, 1961.

Platt, John R. "Are Bats Facing a Hidden Extinction Crisis?" *Scientific American,* August 12, 2016. https://blogs.scientificamerican.com/extinction-countdown/bats-hidden-extinction/.

---. "Snails Are Going Extinct: Here's Why That Matters." *Scientific American,* August 10, 2016. https://blogs.scientificamerican.com/extinction-countdown/snails-going-extinct/.

Poore, J., and T. Nemicek. "Reducing food's environmental impacts through producers and consumers." *Science* 360, no. 6392 (June 1, 2018): 987–992. https://doi.org/10.1126/science.aaq0216.

Post-Carbon Institute (website). "Relocalize." n.d. Accessed March 1, 2019. https://www.postcarbon.org/relocalize/.

Popper, Deborah, and Frank Popper. "The Great Plains: From Dust to Dust." *Planning* 53 (1987): 12–18. https://www.cpp.edu/~tgyoung/rs510/Grt_Plains.pdf.

Poynter, Jayne. *The Human Experiment: Two Years and Twenty Minutes Inside Biosphere 2.* Thunder's Mouth Press, 2006.

Provan, Iain. *Convenient Myths: The Axial Age, Dark Green Religion, and the World That Never Was.* Baylor University Press, 2013.

Pugh, Tom. "Are young trees or old forests more important for slowing climate change?" The Conversation (website), July 30, 2020. https://theconversation.com/are-young-trees-or-old-forests-more-important-for-slowing-climate-change-139813.

Pugh, Thomas A. M., Mats Lindeskog, Benjamin Smith, Benjamin Poulter, Almut Arneth, Vanessa Haverd, and Leonardo Calle. "Role of forest regrowth in global carbon sink dynamics." *Proceedings of the National Academy of Science* 116, no. 10 (March 5, 2019): 4382–4387. https://doi.org/10.1073/pnas.1810512116.

Putnam, Robert D. *Bowling Alone: The Collapse and Revival of American Community* Touchstone, 1995.

Radiation Effects Research Foundation. "Frequently asked questions," n.d., accessed August 20, 2020. https://www.rerf.or.jp/en/faq/.

Ramankutty, Navin, Amato T. Evan, Chad Monfreda, and Jonathan A. Foley. "Farming the planet: 1. Geographic distribution of global agricultural lands in the year 2000." *Global Biogeochemical Cycles* 22, GB1003 (January 17, 2008). https://doi.org/10.1029/2007GB002952.

Ramankutty, Navin, and Jonathan A. Foley. "Estimating historical changes in global land cover: croplands from 1700 to 1992." *Global Biogeochemical Cycles* 13, no. 4 (December 1999): 997–1027. https://doi.org/10.1029/1999GB900046.

Rao, Sailesh K., Atul K. Jain, and Shijie Shu. "The Lifestyle Carbon Dividend: Assessment of the Carbon Sequestration Potential of Grasslands and Pasturelands Reverted to Native Forests" (poster session). AGU Fall Meeting, December 14–18, 2015. https://agu.confex.com/agu/fm15/meetingapp.cgi/Paper/67429.

Rapp, Anders, H. N. Le Houérou, and Bengt Lundholm, editors. *Can Desert Encroachment be Stopped? A Study with Emphasis on Africa.* Swedish Natural Science Research Council, 1976.

Raworth, Kate. *Doughnut Economics: Seven Ways to Think Like a 21st Century Economist.* Chelsea Green Publishing, 2017.

Razo-León, Alvaro E., Miguel Vásquez-Bolaños, Alejandro Muñoz-Urias, and Francisco M. Huerta-Martínez. "Changes in bee community structure (Hymenoptera, Apoidea) under three different land-use conditions." *Journal of Hymenoptera Research* 66 (October 31, 2018): 23–38. https://doi.org/10.3897/jhr.66.27367.

Reed, Stanley. "Nuclear Fusion Edges Toward the Mainstream," *New York Times*, October 18, 2021. https://www.nytimes.com/2021/10/18/business/fusion-energy.html.

Reynolds, George. "Why do people hate vegans?" *The Guardian*, October 25, 2019. https://www.theguardian.com/lifeandstyle/2019/oct/25/why-do-people-hate-vegans.

Reynolds, Mike. "Saskatchewan: The birthplace of Passive House and passive solar home design." Ecohome (website), October 5, 2020. https://www.ecohome.net/guides/1422/passive-house-saskatchewan-the-birthplace-of-high-performance-buildings-and-passive-solar-home-design/.

Ricardo, David. *On the Principles of Political Economy and Taxation*, 1821. The Library of Economics and Liberty (website), 1999. https://www.econlib.org/library/Ricardo/ricP.html.

Ritchie, Hannah. "What was the death toll from Chernobyl and Fukushima?" *Our World in Data*, July 24, 2017. https://ourworldindata.org/what-was-the-death-toll-from-chernobyl-and-fukushima/.

Ritchie, Hannah, and Max Roser. "Meat and Dairy Production," revised November 2019. https://ourworldindata.org/meat-production.

Reig, Paul. "What's the Difference Between Water Use and Water Consumption?" World Resources Institute (website), March 12, 2013. https://www.wri.org/blog/2013/03/what-s-difference-between-water-use-and-water-consumption.

Richter, Brian. *Chasing Water: A guide to moving from scarcity to sustainability*. Island Press, 2014.

Rhodes, Joshua D. "The old, dirty, creaky US electric grid would cost $5 trillion to replace." *The Conversation*, March 16, 2017. https://theconversation.com/the-old-dirty-creaky-us-electric-grid-would-cost-5-trillion-to-replace-where-should-infrastructure-spending-go-68290.

Roberts, David. "This climate problem is bigger than cars and much harder to solve." *Vox*, October 10, 2019. https://www.vox.com/energy-and-environment/2019/10/10/20904213/climate-change-steel-cement-industrial-heat-hydrogen-ccs.

Robin, Vicki, and Joe Dominguez. *Your Money or Your Life: 9 Steps to Transforming Your Relationship with Money and Achieving Financial Independence*. Viking, 1992.

Romm, Joe. "How the EPA and New York Times Are Getting Methane All Wrong." *ThinkProgress*, August 20, 2015. https://archive.thinkprogress.org/how-the-epa-and-new-york-times-are-getting-methane-all-wrong-eba3397ce9e5/.

Rosegrant, Mark W., Ximing Cai, and Sarah A. Cline. *World Water and Food to 2025*. International Food Policy Research Institute, Washington, DC, 2002. https://reliefweb.int/sites/reliefweb.int/files/resources/1B0BBA6D1080C010C1256C6E002E1D17-ifpri-water2025-16oct.pdf.

Ruddiman, William F. *Plows, Plagues, and Petroleum: How Humans Took Control of the Climate*. Princeton University Press, 2010.

Sánchez-Bayo, Francisco, and Kris A. G. Wyckhuys. "Worldwide decline of the entomofauna: A review of its drivers." *Biological Conservation* 232 (2019): 8–27, ISSN 0006-3207, https://doi.org/10.1016/j.biocon.2019.01.020.

Sasaki, Fumio. *Goodbye, Things: The New Japanese Minimalism.* Translated by Eriko Sugita. W. W. Norton, 2017.

Saunders, Caroline, Andrew Barber, and Greg Taylor. "Food Miles—Comparative Energy/Emissions Performance of New Zealand's Agriculture Industry." Research Report No. 285 (2006). Agribusiness and Economics Research Unit, Lincoln University, New Zealand. https://researcharchive.lincoln.ac.nz/bitstream/handle/10182/125/aeru_rr_285.pdf.

Scanlon, Bridget R., Claudia C. Faunt, Laurent Longuevergne, Robert C. Reedy, William M. Alley, Virginia L. McGuire, and Peter B. McMahon. "Groundwater depletion and sustainability of irrigation in the US High Plains and Central Valley." *Proceedings of the National Academy of Science*, 109, no. 24 (June 2012): 9320–9325. https://doi.org/10.1073/pnas.1200311109.

Schlosser, Eric. "Asne Seierstad's 'One of Us,' About Rampage in Norway." *New York Times Review of Books*, April 20, 2015. https://www.nytimes.com/2015/04/26/books/review/asne-seierstads-one-of-us-about-rampage-in-norway.html.

Schmit, Julie. "In California, Demand for Groundwater Causing Huge Swaths of Land to Sink." National Geographic (website), March 26, 2014. https://www.nationalgeographic.com/news/2014/3/140325-california-drought-subsidence-groundwater/.

Schneider-Mayerson, Matthew. *Peak Oil: Apocalyptic Environmentalism and Libertarian Political Culture.* University of Chicago Press, 2016.

Schor, Juliet B. *The Overworked American: The Unexpected Decline of Leisure.* Basic Books, 1991.

---. *Plenitude: the new economics of true wealth.* Penguin Press, 2010.

Schumacher, Ernst F. *Small Is Beautiful: Economics as if People Mattered.* Perennial Library / Harper & Row, 1975.

Scully, Matthew. *Dominion: The Power of Man, the Suffering of Animals, and the Call to Mercy.* St. Martin's Griffin, 2003.

Searchinger, Timothy D., Stefan Wirsenius, Tim Beringer, and Patrice Dumas. "Assessing the efficiency of changes in land use for mitigating climate change," *Nature* 564 (December 13, 2018): 249–253. https://doi.org/10.1038/s41586-018-0757-z.

Sennett, Richard. *The Corrosion of Character: The Personal Consequences of Work in the New Capitalism.* W. W. Norton and Company, Inc., 1998.

Shah, Tushaar, David Molden, R. Sakthivadivel, and David Seckler. "The Global Groundwater Situation: Overview of Opportunities and Challenges." Colombo, Sri Lanka: International Water Management Institute, 2000. https://core.ac.uk/download/pdf/6472688.pdf.

Shenker, Barry. *Intentional Communities: Ideology and Alienation in Communal Societies.* Routledge & Kegan Paul, 1986.

Shiklomonov, Igor. *World Water Resources: a New Appraisal and Assessment for the 21st Century.* UNESCO, 1998. https://www.caee.utexas.edu/prof/mckinney/ce385d/Papers/Shiklomanov.pdf.

Singer, Peter. *Animal Liberation: A New Ethics for Our Treatment of Animals.* Avon Books, 1977.

---. *The Expanding Circle: Ethics and Sociobiology.* Farrar, Strauss, & Giroux, 1981.

Skidelsky, Edward. "Unconditional Basic Income and Degrowth keynote lecture." The Unconditional Basic Income and Degrowth conference, Hamburg, Germany, May 19–20, 2016. Video of lecture (length: 17:33). https://basicincome.org/news/2016/06/video-edward-skidelsky-unconditional-basic-income-and-degrowth-keynote-lecture/.

Skidelsky, Robert, and Vijay Joshi. "Keynes, Global Imbalances, and International Monetary Reform, Today." *Robert Skidelsky* (blog), June 23, 2010. https://robertskidelsky.com/2010/06/23/keynes-global-imbalances-and-international-monetary-reform-today/.

Smil, Vaclav. "Energy in the Twentieth Century: Resources, Conversions, Costs, Uses, and Consequences," *Annu. Rev. Energy Environ.* 25 (November 2000): 21–51. https://doi.org/10.1146/annurev.energy.25.1.21.

---. *Enriching the Earth: Fritz Haber, Carl Bosch, and the Transformation of World Food Production*. MIT Press, 2001.

---. "Harvesting the Biosphere," *Population and Development Review* 37, no. 4 (December 2011): 613–636. http://vaclavsmil.com/wp-content/uploads/PDR37-4.Smil_.pgs613–636.pdf.

---. *Harvesting the Biosphere: What We Have Taken from Nature*. MIT Press, 2013.

---. "Nitrogen and food production: Proteins for human diets." *Ambio* 31, no. 2 (March 2002): 126–131. http://vaclavsmil.com/wp-content/uploads/docs/smil-article-2002-nitrogen-and-food-production.pdf.

---. "Nitrogen cycle and world food production." *World Agriculture* 2 (2011): 9–13. http://vaclavsmil.com/wp-content/uploads/docs/smil-article-*worldagriculture.pdf*.

---. *Should We Eat Meat?*: Evolution and Consequences of Modern Carnivory. John Wiley and Sons, 2013.

Smith, Felisa, Scott Elliott, and S. Lyons. "Methane emissions from extinct megafauna." *Nature Geoscience* 3, 374–375 (2010). 10.1038/ngeo877.

Squires, David, and Chloe Anderson, "U.S. Health Care from a Global Perspective: Spending, Use of Services, Prices, and Health in 13 Countries." The Commonwealth Fund 15 (October 2015). https://www.commonwealthfund.org/sites/default/files/documents/___media_files_publications_issue_brief_2015_oct_1819_squires_us_hlt_care_global_perspective_oecd_intl_brief_v3.pdf.

Staalesen, Atle. "New Sinkholes Appear in Yamal." *Barents Observer*, February 12, 2015. https://barentsobserver.com/en/arctic/2015/02/new-sinkholes-appear-yamal-12-02.

Statistica (website). "Global oil production." Accessed June 20, 2018. https://www.statista.com/statistics/265229/global-oil-production-in-million-metric-tons/.

Steinhart, J. S., and C. E. Steinhart. "Energy Use in the U. S. Food System." *Science* 184, no. 4134 (April 19, 1974): 307–316. https://doi.org/10.1126/science.184.4134.307.

Strassburg, Bernardo B. N., Alvaro Iribarrem, Hawthorne L. Beyer, Carlos Leandro Cordeiro, et al. "Global priority areas for ecosystem restoration." *Nature* 586 (October 14, 2020): 724–729. https://doi.org/10.1038/s41586-020-2784-9.

Sullivan, J. L., C. E. Clark, J. Han, and M. Wang. "Life Cycle Analysis Results of Geothermal Systems in Comparison to Other Power Systems." August 2010. Argonne National Laboratory, Energy Systems Division, Center for Transportation Research. https://www.energy.gov/eere/geothermal/downloads/life-cycle-analysis-results-geothermal-systems-comparison-other-power.

Sun Tzu. *The Art of War.* Translated by Thomas Cleary. Shambhala Publications, 2005.

Swanson, Heather Anne. "The Banality of the Anthropocene." Society for Cultural Anthropology, Member Voices, *Fieldsights*, February 22, 2017. https://culanth.org/fieldsights/the-banality-of-the-anthropocene.

Swenson, Shea. "What Can Be Done About the Phosphorus Crisis?" *Modern Farmer*, May 4, 2022. https://modernfarmer.com/2022/05/phosphorus-fertilizer-crisis/.

Tabary, Zoe. "Scientists caught off-guard by record temperatures linked to climate change." Thompson Reuters Foundation, July 26, 2016. https://news.trust.org/item/20160726133558-a7f25.

Tainter, Joseph. *The Collapse of Complex Societies.* Cambridge University Press, 1988.

Taubenberger, Jeffery K., and David M. Morens. "1918 Influenza: the mother of all pandemics." *Emerging infectious diseases* 12, no.1 (2006): 15–22. doi:10.3201/eid1201.050979.

Thorium MSR Foundation (website). "In Depth: Clean Industrial Heat," n.d., accessed May 18, 2020. https://www.thmsr.com/en/heat/.

Thornton, Philip K. "Livestock production: recent trends, future prospects," *Philos Trans R Soc Lond B Biol Sci.* 365, no. 1554 (September 27, 2010): 2853–2867. https://doi.org/10.1098/rstb.2010.0134.

Troeh, Frederick R., J. Arthur Hobbs, and Roy L. Donahue. *Soil and Water Conservation: For Productivity and Environmental Protection.* Prentice Hall, 2004.

Troeh, Frederick R., and Louis Thompson, *Soils and Soil Fertility*, 6th edition. Blackwell Publishing, 2005.

Turchin, Peter. *Ages of Discord: A Structural-Demographic Analysis of American History.* Beresta Books, 2016.

---. "Collapse of Complex Societies: Did Drought Kill off the Mayans?" *Cliodynamica* (blog), November 11, 2012. https://peterturchin.com/cliodynamica/collapse-of-complex-societies-did-drought-kill-off-the-mayans/.

---. "The Ginkgo Model of Societal Crisis," *Cliodynamica* (blog), August 16, 2018. https://peterturchin.com/cliodynamica/the-ginkgo-model-of-societal-crisis/.

---. "A History of the Near Future: What history tells us about our Age of Discord." *Cliodynamica*, November 6, 2019. Abstract posted at: http://peterturchin.com/cliodynamica/a-history-of-the-near-future-what-history-tells-us-about-our-age-of-discord/. Slides for talk posted at: https://peterturchin.com/wp-content/uploads/2019/11/MPF2019.pdf.

---. "Political instability may be a contributor in the coming decade." *Nature* 463 (February 4, 2010), 608.

---. "The Rise and Demise of Complex Societies. The Effect of Climate. Science," *Cliodynamica* (blog), November 4, 2012. https://peterturchin.com/cliodynamica/the-rise-and-demise-of-complex-societies-the-effect-of-climate-science/.

---. *Ultrasociety: How 10,000 Years of War Made Humans the Greatest Cooperators on Earth.* Beresta Books, 2018.

Turchin, Peter, and Sergey Nefedov. *Secular Cycles.* Princeton University Press, 2009.

Tverberg, Gail. "The climate change story is half true." *Our Finite World*, April 30, 2019. https://ourfiniteworld.com/2019/04/30/the-climate-change-story-is-half-true/.

---. "Deflationary Collapse Ahead?" *Our Finite World*, August 26, 2015. https://ourfiniteworld.com/2015/08/26/deflationary-collapse-ahead/.

---. "How the Economy Works as It Reaches Energy Limits—An Introduction for Actuaries and Others." *Our Finite World*, May 11, 2018. https://ourfiniteworld.com/2018/05/11/how-the-economy-works-as-it-reaches-energy-limits-an-introduction-for-actuaries-and-others/.

---. "A new theory of energy and the economy—Part 1—Generating economic growth." *Our Finite World*, January 21, 2015. https://ourfiniteworld.com/2015/01/21/a-new-theory-of-energy-and-the-economy-part-1-generating-economic-growth/.

United Nations Department of Economic and Social Affairs, Population Division. *World Population Prospects: The 2015 Revision, Key Findings and Advance Tables*. New York, May 2015. https://population.un.org/wpp/Publications/Files/Key_Findings_WPP_2015.pdf.

---. *World Population Prospects 2022: Summary of Results*. New York, 2022. https://www.un.org/development/desa/pd/content/World-Population-Prospects-2022.

United Nations Development Program. "Human Development Report 2014: Sustaining Human Progress – Reducing Vulnerabilities and Building Resilience." New York, 2014. http://hdr.undp.org/en/content/human-development-report-2014.

United States Department of Agriculture. "Crop Production Historical Track Records." ISSN: 2157-8990. April 2018. https://www.nass.usda.gov/Publications/Todays_Reports/reports/croptr18.pdf.

---. *2012 National Resources Inventory Summary Report*. August 2015. https://www.nrcs.usda.gov/Internet/FSE_DOCUMENTS/nrcseprd396218.pdf.

United States Department of Labor. "Coal Fatalities for 1900 Through 2018," n.d. (2019?), accessed March 5, 2019. https://arlweb.msha.gov/stats/centurystats/coalstats.asp. This site is periodically updated with the data and the title changing slightly.

---. "MSHA issues final rule on lowering miners' exposure to respirable coal dust," press release, April 23, 2014. https://www.dol.gov/newsroom/releases/msha/msha20140669.

United States Energy Information Administration. "Monthly Energy Review." January 2017. http://large.stanford.edu/courses/2017/ph241/sheu1/docs/eia-0035-1-2017.pdf.

---. "Petroleum and other liquids." Periodically updated, accessed February 9, 2019. https://www.eia.gov/dnav/pet/hist/RBRTED.htm.

United States Geological Survey. "Land Subsidence in the United States." USGS Fact Sheet-165-00. December 2000. Credit: Dr. Joseph F. Poland, USGS. Public domain. https://water.usgs.gov/ogw/pubs/fs00165/SubsidenceFS.v7.PDF.

University of Adelaide. "Humans hastened the extinction of the woolly mammoth." *ScienceDaily*, November 11, 2021. https://www.sciencedaily.com/releases/2021/11/211111130304.htm.

Utrecht University Faculty of Science. "Methane promising route for storage of renewable energy from sun and wind." *Phys.Org*, April 1, 2019. https://phys.org/news/2019-04-methane-route-storage-renewable-energy.html.

Vail, Jeff. "The Renewables Gap: The Political Challenge of Affecting a Societal Transition to Renewable Sources of Energy." ASPO 2009 International Peak Oil Conference, Denver, Colorado, October 10, 2009. http://theoildrum.com/node/5965.

Van Boeckel, Thomas P., Charles Brower, Marius Gilbert, Bryan T. Grenfell, et al. "Global trends in antimicrobial use in food animals." *Proceedings of the National Academy of*

Varnum, Michael E. W., and Igor Grossmann, "Cultural Change: The How and the Why." *Perspectives on Psychological Science* 12, no. 6 (September 15, 2017): 956–972. https://doi.org/10.1177/1745691617699971.

Vegan Society, The. "How to go vegan." https://www.vegansociety.com/go-vegan/how-go-vegan.

Vettese, Troy, and Drew Pendergrass. Half-Earth Socialism: A Plan to Save the Future from Extinction, Climate Change, and Pandemics. Verso, 2022.

Vidal, John. "Cut world population and redistribute resources, expert urges." *The Guardian*, April 26, 2012. https://www.theguardian.com/environment/2012/apr/26/world-population-resources-paul-ehrlich.

Vikström, Hanna, Simon Davidsson, and Mikael Höök. "Lithium availability and future production outlooks." *Applied Energy* 110 (October 2013): 252–266. https://doi.org/10.1016/j.apenergy.2013.04.005.

Villholth, Karen G. "Comprehensive Assessment of Consequences and Options for Agricultural Groundwater Use." International Water Management Institute. Paper prepared for the workshop: "Groundwater Governance and Management - An IWRM Perspective in Arid / Semi-Arid Climates," Cairo, Egypt, April 4–7, 2005. http://www.iwmi.org/Assessment/files/Synthesis/Groundwater/karen_Paper_Comprehensive%20Assessment_Agricultural%20Gw%20Use_Cairo.doc.

Vitousek, Peter M., Paul R. Ehrlich, Anne H. Ehrlich, and Pamela A. Matson. "Human Appropriation of the Products of Photosynthesis." *Bioscience* 36, no. 368 (June 1986): 368–373. https://doi.org/10.2307/1310258.

Vogel, Gretchen. "Where have all the insects gone?" *Science*, May 10, 2017. https://doi.org/10.1126/science.aal1160.

Vogt, Charlotte, Matteo Monai, Gert Jan Kramer, and Bert M. Weckhuysen. "The renaissance of the Sabatier reaction and its applications on Earth and in space." *Nat Catal* 2, 188–197 (2019). https://doi.org/10.1038/s41929-019-0244-4.

Waldrop, M. Mitchell. "Nuclear goes retro—with a much greener outlook." *Knowable Magazine*, February 2, 2019. https://doi.org/10.1146/knowable-022219-2.

Wallace-Wells, David. "The Uninhabitable Earth." *New York* magazine, July 10, 2017. https://nymag.com/intelligencer/2017/07/climate-change-earth-too-hot-for-humans.html.

---. *The Uninhabitable Earth: Life After Warming.* Tim Duggan Books, 2019.

Warren, A., and J. K. Maizels, "Ecological Change and Desertification." In *Desertification: Its Causes and Consequences*, United Nations Secretariat Staff, Pergamon Press, 1977.

Watts, Jonathan. "We have 12 years left to limit climate catastrophe, warns UN." *The Guardian*, October 8, 2018. https://www.theguardian.com/environment/2018/oct/08/global-warming-must-not-exceed-15c-warns-landmark-un-report.

Webster, Robert G. "Wet markets—a continuing source of severe acute respiratory syndrome and influenza?" Lancet 363, no. 9404 (2004): 234–6. doi:10.1016/S0140-6736(03)15329-9.

Weis, Tony. *The Ecological Hoofprint: The Global Burden of Industrial Livestock.* Zed Books, 2013.

Wenger, Jeffrey B. "Working for $7.25 an Hour: Exploring the Minimum Wage Debate." The Rand Blog, September 1, 2016. https://www.rand.org/blog/2016/09/working-for-725-an-hour-exploring-the-minimum-wage.html.

Wetterer, J. K., C. Dunning, M. Yospin, and A. Himler. "Invertebrate Diversity and Ecology in Biosphere 2." *Sonoran Arthropod Studies Institute*, Invertebrates in Captivity Conference Proceedings. Tucson, AZ, 1997, 119–122. https://www.researchgate.net/publication/256979633_Invertebrate_diversity_and_ecology_in_Biosphere_2.

Wetterer, J. K., S. E. Miller, D. E. Wheeler, C. A. Olson, et al. "Ecological Dominance by Paratrechina Longicornis (hymenoptera: formicidae), an Invasive Tramp Ant, in Biosphere 2." *The Florida Entomologist* 82 (September 1999): 381–388. https://doi.org/10.2307/3496865.

Wilkenson, Richard, and Kate Pickett. *The Spirit Level: How Greater Equality Makes Societies Stronger.* Bloomsbury Press, 2009.

Williams, Florence. "Plains Sense." *High Country News*, January 15, 2001. https://www.hcn.org/issues/194/great-plains-sense-buffalo.

Williams, Michael. "Forests." In *The Earth as Transformed by Human Action*, B. L. Turner II, William C. Clark, Robert W. Kates, John F. Richards, Jessica T. Mathews, and William B. Meyer, eds. (Cambridge University Press, New York, 1990), p. 179–201.

Williams, Stephen. "How Molten Salt Reactors Might Spell a Nuclear Energy Revolution." *ZME Science*, updated June 4, 2020. https://www.zmescience.com/science/what-is-molten-salt-reactor-424343/.

Williams, Stephen J., and Rod Taylor, editors. *Sustainability and the New Economics: Synthesising Ecological Economics and Modern Monetary Theory*. Springer, 2022.

Wilson, Edward O. *Biophilia*. Harvard University Press, 1984.

---. *Half-Earth: Our Planet's Fight for Life*. Liveright Publishing Company, 2016.

---. "The loss of biodiversity is a tragedy." UNESCO Media Services (website), February 9, 2010. http://www.unesco.org/new/en/media-services/single-view/news/edward_o_wilson_the_loss_of_biodiversity_is_a_tragedy/.

Wilson, S. S. "Bicycle Technology." *Scientific American*, March 1973, 81–91. https://www.jstor.org/stable/24923004.

Wood, Graeme. "The Next Decade Could be Even Worse," *The Atlantic*, December 2020. https://www.theatlantic.com/magazine/archive/2020/12/can-history-predict-future/616993/.

World Health Organization. "The World Health Report 1996." (The online version of this report at https://apps.who.int/iris/handle/10665/60924 lacks some pages which I cite.)

---. "The World Health Report 2007." https://apps.who.int/iris/handle/10665/43713.

World Nuclear Association (website). "What is nuclear waste, and what do we do with it?" n.d., accessed June 9, 2022. https://www.world-nuclear.org/nuclear-essentials/what-is-nuclear-waste-and-what-do-we-do-with-it.aspx.

Wright, Erik Olin. *Envisioning Real Utopias*. Verso, 2010.

Wright, Ronald. *A Short History of Progress*. House of Anansi Press, 2004.

Yergin, Daniel. "It's not the end of the oil age," *Washington Post*, July 31, 2005. https://www.washingtonpost.com/wp-dyn/content/article/2005/07/29/AR2005072901672.html.

Young, Anthony. *Land Resources: Now and for the future*. Cambridge University Press, 1998.

Zablocki, Benjamin. *Alienation and Charisma: A Study of Contemporary American Communes*. Free Press, 1980.

Zaraska, Marta. *Meathooked: The History and Science of our 2.5-Million-Year Obsession with Meat*. Basic Books, 2016.

Zhang, Tao, Qunfu Wu, and Zhigang Zhang. "Probable Pangolin Origin of SARS-CoV-2 Associated with the COVID-19 Outbreak." *Current Biology* 30 (April 6, 2020): 1346–1351. https://doi.org/10.1016/j.cub.2020.03.022.

INDEX

www.ingramcontent.com/pod-product-compliance
Lightning Source LLC
Chambersburg PA
CBHW060022030426
42334CB00019B/2136